The Cambridge Manuals of Science and
Literature

THE MAKING OF LEATHER

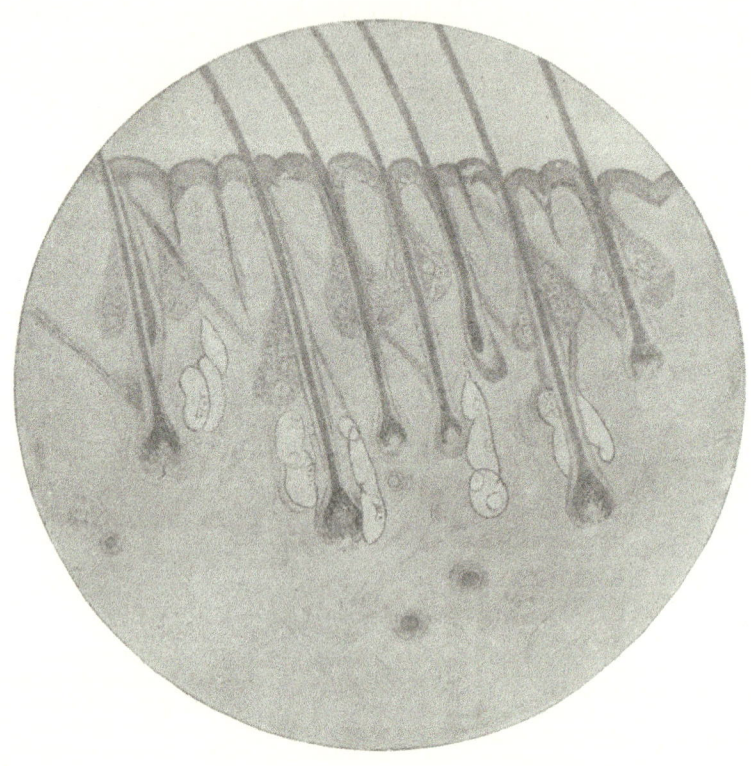

Fig. 1. Diagram Section of Calf Skin.

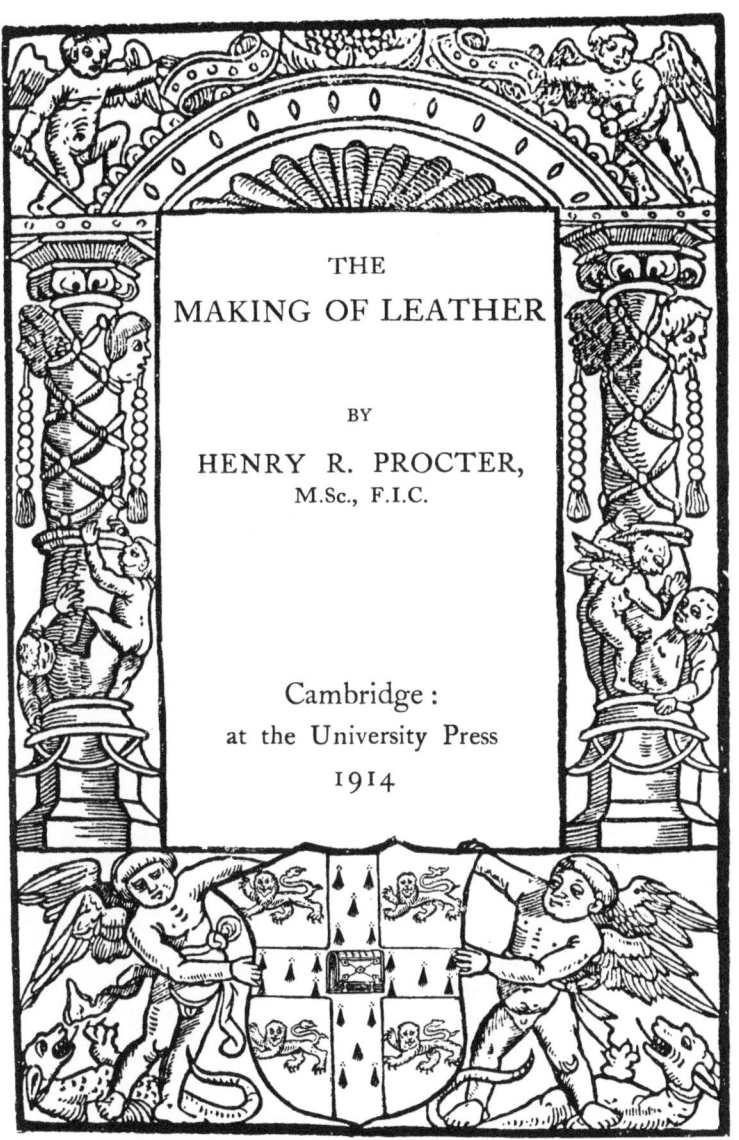

THE
MAKING OF LEATHER

BY

HENRY R. PROCTER,
M.Sc., F.I.C.

Cambridge:
at the University Press
1914

CAMBRIDGE UNIVERSITY PRESS
Cambridge, New York, Melbourne, Madrid, Cape Town,
Singapore, São Paulo, Delhi, Tokyo, Mexico City

Cambridge University Press
The Edinburgh Building, Cambridge CB2 8RU, UK

Published in the United States of America by
Cambridge University Press, New York

www.cambridge.org
Information on this title: www.cambridge.org/9781107401846

First published 1914
First paperback edition 2011

A catalogue record for this publication is available from the British Library

ISBN 978-1-107-40184-6 Paperback

*With the exception of the coat of arms
at the foot, the design on the title page is a
reproduction of one used by the earliest known
Cambridge printer, John Siberch,* 1521

PREFACE

THE present book must in no sense be taken as a manual of the art of leather manufacture, and for many reasons working details have been purposely avoided except where they seemed necessary to the understanding of principles. Those who are concerned in actual manufacture will of course consult the larger works mentioned in the bibliography on p. 148; but there are probably readers who will be satisfied with a sketch of the methods and some discussion of the scientific basis of a very ancient and important industry which, nevertheless, involves in its explanation some of the most difficult branches of modern knowledge; and which, from this cause, has lagged behind most others in its scientific development. To realise these difficulties, it is only necessary to remember that not only is the skin a complex anatomical structure, but that its constituents as well as most tanning materials, belong to the class of uncrystallizable substances known as "colloids," which are not only usually impossible to separate in pure form, but of which the modes of reaction differ so markedly from those of the simpler crystallizable substances that their study has practically formed a new branch of chemistry.

The reader must therefore not complain if the ground covered is somewhat unfamiliar, but it is well worth exploring, since the explanation of some of the greatest problems of physiology and life undoubtedly lie in the same region.

For illustrations of leather working machinery the author is indebted to: Messrs The Turner Co., B. & D. Wright & Sons Ltd., E. Wilson & Sons, and Leidgen Machine Co.

HENRY R. PROCTER.

May 1914

CONTENTS

CHAP.   PAGE

PREFACE vii

I. INTRODUCTORY 1

II. HIDES AND SKINS 2

III. CURING OF HIDES AND SKINS 5

IV. THE STRUCTURE OF THE SKIN 9

V. THE CHEMISTRY OF THE SKIN 16

VI. THE PRELIMINARY PROCESSES. SOAKING . . 22

VII. UNHAIRING 24

VIII. CHEMICAL DELIMING 37

IX. BACTERIA AND FERMENTATION 52

X. THE FERMENTIVE "BATES" 55

XI. THE CONVERSION OF SKIN INTO LEATHER . . 67

XII. THE PICKLING PROCESS 72

XIII. ALUMED LEATHER 80

XIV. THE BASIC CHROME PROCESS 88

XV. THE TWO-BATH CHROME PROCESS . . . 93

XVI. THE VEGETABLE TANNING MATERIALS . . 99

XVII. THE VEGETABLE TANNING PROCESS . . . 104

XVIII. CURRIED LEATHERS 117

XIX. MOROCCOS AND FANCY LEATHERS . . . 128

XX. OIL LEATHERS 134

XXI. THE USE AND CARE OF LEATHER . . . 140

BIBLIOGRAPHY 148

INDEX 150

LIST OF ILLUSTRATIONS

FIG.   PAGE

1. Diagram Section of Calf Skin   .   *Frontispiece*

2. Unhairing Knife   34

3. Fleshing Knife   34

4. Tanners' Beam   34

5. Leidgen's Unhairing Machine   .   .   .   36

6. Fleshing Machine. Front view. Turner Co.   38

7. Fleshing Machine. Back view. Turner Co.   39

8. Paddle Vat   57

9. Wright's Glazing Machine   98

10. Striking Pin   114

11. Wilson's Butt Rolling Machine .   .   .   116

12. Wilson's Striking Machine   116

13. Slicker   122

14. Curriers' Shaving Knife   123

15. Curriers' Steels   123

CHAPTER I

INTRODUCTORY

IN writing of a manufacture so ancient as that of leather, it is difficult to begin without some sort of historical notice ; and yet harder to do so with success, since in the earliest historical times the art had attained such a development that its details were no longer a matter of curiosity, and hence little information of its methods has been preserved. We know, from actual specimens, that in Rome, presumably in Greece, and certainly in the still earlier civilisation of Egypt, leather was used for most of the purposes for which we use it to-day.

Looking yet further behind us we may picture the ways of the primitive hunter from those of savage races who still use skins for clothing, and prepare them in the old traditional ways of their tribe. We may imagine that the earliest of these depended on greasing, smoking, stretching and softening the skin as it dried ; that only later was the use of barks and berries discovered, and later still that of alum. These speculations serve little purpose except to show that while methods and machines develop, principles remain unchanged ; and all the primitive types of which we have spoken survive in altered forms in the manufacture of to-day, alongside others of which our ancestors knew nothing.

CHAPTER II

HIDES AND SKINS

In those times when man lived mainly on the products of the chase, the skins of wild animals formed the principal source of leather, while their place is now quite a subordinate one, except for furs, of which the present book does not treat further than regards the conversion of the skins into a permanent and imputrescible material, which is strictly a species of leather. Apart from fur-bearing animals, deer and antelopes have some importance in the production of oil leathers related in manufacture to so-called "chamois," and principally used for winter gloves. The supply of these skins is a somewhat irregular one, if we except those of the reindeer, which is now not merely domesticated in the far

North, but kept in a semi-wild state in the mountain
regions of Norway, where immense herds wander
from place to place under the control of a few Lapps
and their dogs, picking up a subsistence in summer
on the lichens and scanty vegetation of the "high
fells" above the levels (say 4000 ft.) which are reached
by the cattle of the saeters. "Chamois" or "wash-
leather" was no doubt once the skin of the animal
with which its name connects it, and therefore
belonged to the class of which we are speaking, but
it is now simply the inner half of sheepskin, and will
be discussed in its proper place. In passing, it may
be mentioned that "dogskin," and many imitations of
crocodile, pig and goat, are derived from the same
harmless animal.

Among the still important wild skins are those
of some marine animals, and specially of the arctic
seals of various species, which differ from the Alaskan
fur seal in having much coarser coats, but which
are captured in large quantities for their oil and skins.

The skins of alligators are the source of genuine
"crocodile" hide, lizard skins are used for purses and
fancy articles, while against snake skins there exists
a feminine prejudice, either from historical reasons,
or because they are only smooth if stroked the right
way. Fish skins, except those of the shark tribe, are
little used, though some of them are very suitable for
small fancy articles.

By far the most important materials for leather manufacture are furnished by domestic animals, and especially by the ox and sheep.

It is an important feature of the industry that its main raw material is a by-product of our food supply, and no animals are grown specially for their skins ; and hence demand does not in this case have any perceptible effect on supply, which, with the increased growth of corn, and the consequent diminution of prairie lands, constantly tends to become short of requirements, and consequently dearer.

As a rule the wildest cattle, and those more exposed to natural climatic conditions, have the thickest hides, while breeding directed mainly to increase meat and milk tends to a thinner and larger hide of finer texture.

Very similar considerations apply to sheep, in which the desire both for size and for finer or longer wool has affected the character of the skin ; which is also less mature than formerly as the sheep are killed much younger.

Sex and age have as much influence on the character of the skin as they have on other bodily developments. Cow, and especially heifer, hides are thin and fine, while bull hides are loose in texture, thin on the back and very coarse and thick on the under part of the body and on the head. Ox hides have a quite different character ; the hide is the thickest on the back, and especially over the kidneys, and thin on the belly and head, and its texture is finer and much more compact. It forms the principal material for sole-leather.

Of foreign domestic animals, the small Indian cattle, and the oriental buffalo must be specially mentioned. The hides of the former are known in Europe as "kips," probably from their diminutive size, as a "kip" is strictly the skin of a young beast older than a calf; while Indian kips are full-grown.

The oriental buffalo, which must not be confused with the now almost extinct bison of the American plains, is a large and often almost hairless animal, much used for draught and other domestic purposes throughout Asia, and even in the south-eastern portions of Europe. The hide is large, thick, and coarse, and somewhat loose in texture, so that it is mostly used as a cheap and inferior sole-leather.

CHAPTER III

'CURING' OF HIDES AND SKINS

THE hides and skins of the English butchery are weekly sold by auction at the various hide markets scattered throughout the country : though there are still a few tanners and fellmongers in country districts who collect direct from the butchers. As only a few days should elapse from the slaughter of the animal to the beginning of the tanner's work, no special means, beyond slight salting in warm weather, are required to prevent putrefaction.

The case is very different with imported hides and skins, which in many cases must cross the tropics, and with these the method of preservation adopted has an effect on their quality and subsequent treatment almost as important as that of the breed of the animal. On the whole the most satisfactory method in common use is that of salting; but to be effective this must be of a different character to the slight sprinkling which suffices for English hides. As an example of the best method, that employed in the Chicago stockyards may be described; this differs in no important respect from those used on the River Plate and other large exporting centres.

The hides, as soon as cool after removal from the animal, are laid flat and hairside downwards, in large cool cellars so as to cover a space of 15 or 20 ft. square, and the edges of the outer hides, after being freely sprinkled with salt, are turned in, not only to make the space a regular rectangle, but to raise slightly the outer edges of the pile of which it is to form the base, so that the salt brine may not run out, but must percolate through the hides. Each layer of hides as it is laid down is freely sprinkled with salt by a "salt-thrower" with a suitable shovel, about 25 $°/_0$ on the hide weight being given. What is used is preferably coarse crystallised manufactured salt, crushed rock-salt being unsatisfactory from the amount of iron and calcium sulphate usually present, which contribute to markings on the hide, generally known as "salt-stains." The process is repeated till the height of the pile approaches the limit over which the thrower can spread the salt, when it is levelled and covered with a layer of salt. In this pile the goods lie for two weeks or more, when they should be completely penetrated with saturated salt solution and considerably dried by the absorption of liquid by the salt and the drainage of the brine.

In hot countries distant from the sea, the cost of salt and the weight of wet hides in transit are often a bar to its use, and hides are preserved simply by drying. If this is done in the shade, and the climate is such that drying is rapid, little is to be urged against it; but if hides are dried slowly or in the sun, serious damage often ensues, which is the more vexatious because it cannot be detected till the hides are again softened. During the first stages of sun-drying, the rapid evaporation keeps down the temperature and little harm results, but as the surface hardens the evaporation of the internal moisture is checked, and the heat attained under a tropical sun,

aided as it often is by incipient putrefaction, may
easily be sufficient to melt the still moist interior
to glue, which dissolves in the liming process, leaving
a blister which frequently develops to a hole.

To lessen this danger, at least in regard to putrefaction, a light preliminary salting is often given
before drying, and such hides are known as "dry-
salted." In India the hides of the small native cattle
usually called "kips" are often cured by scrubbing
the inner or "flesh" side with a soupy mixture of
a natural salt-earth containing mainly sodium sulphate
and carbonate, but little chloride.

Foreign sheep and goat skins are rarely salted,
but usually dried, and are subject to the same sort
of damage as the thicker hides, though to a less
extent, as the drying is easier and quicker. Sheepskins are more valuable for their wool than for the
pelt, and hence most often come to the hands of the
tanner through the "fellmonger," whose main business
it is to remove the wool, and whose treatment of the
skin or "pelt" is usually extremely careless.

CHAPTER IV

THE STRUCTURE OF THE SKIN

WHATEVER the importance of the skin to the tanner, it was still greater to the animal of which it originally formed part ; serving not merely as clothing, but as an organ of feeling and of secretion, and having a complicated structure, to understand which fully we must begin with the beginning of biology.

The first and simplest form of life is that of the single living cell, as it is seen in the Amœba or in the white blood-corpuscle. The white corpuscle is a minute mass of protoplasm or living jelly, existing in millions in the blood, spherical at rest, but capable of motion, forming tentacles (pseudopodia) at will, enveloping with them the bacteria and other minute intruders which it is its business to devour ; and squeezing itself through the minutest pores of the tissues like a semi-liquid mass. In the middle of the jelly is a small oval body different in character from the rest, and called the nucleus, which has important functions connected with the reproduction of the cell. Before this takes place, the nucleus lengthens to an hour-glass form, and finally divides, each half carrying with it a portion of the protoplasm and forming a complete new cell. Such cells are the primary units out of which both plants and animals are built ; but while the blood-corpuscle and the almost identical Amœba are complete individuals, fulfilling in themselves all the requirements of their lives, the higher animals and plants are common-wealths of cells, each of which has its own duties and peculiarities, and each is dependent on the community for its subsistence. It is impossible to say of the unicellular organism whether it is an animal or plant, since the cells of both may consume the same foods, and both may be equally active in move-ment. The real distinction between the higher members of the two kingdoms lies, not in race, but in organisation and government ; the animal organism being much more highly developed, and its govern-ment centralised by a nervous system bringing with it, however, the concomitant evil of vulnerability ; since a serious blow at its centre means sudden and final catastrophe, while in the plant no central government exists.

The history of the individual is more or less an epitome of that of the race. The human embryo is the product of a single cell fertilised by union with a second, for already specialisation has gone so far that neither alone contains all the elements needed for the tremendous effort of evolution about to take place—

the second cell is the detonator without which the dynamite cartridge cannot work. The united cell instantly begins to multiply by division, first into two, and each of these into two again, till in a few days they are millions, united in a single organism. Very early, the process of differentiation begins ; the little disc has an upper and a lower surface and a central part, in each of which the cells differ from the rest, and will take special places in the complete animal. We need not follow the changes through which it passes, till from something lower than the polyp it becomes a breathing conscious vertebrate,— the essentials of its future are already fixed, and time and nourishment only are needed,—but for our present purpose we may remark that the upper layer of cells becomes epidermis, with scales, feathers, hair, horns and nails, according to its kind, while the leather-forming skin, together with bones, ligaments, and muscles, and all the internal framework of the body is derived from the middle layer.

After what has been said, it is almost obvious that the outer coverings of all vertebrates, and *a fortiori* of all mammalia have much in common, which renders a detailed description of each individual skin unnecessary ; and also, that every skin contains two parts of very different origin, and therefore of widely differing character. Of the skins of mammalia it may be said generally that they consist of a thin

outer cellular layer, the *epidermis*, of which hair,
horns and nails are merely specialised products; and
a much thicker inner layer, the *corium*, which is a
felt of gelatinous fibres, the product of cells which
it contains, and which is provided with nerves,
muscles, and blood- and lymph-vessels. Fig. 1
(Frontispiece) shows diagrammatically the arrangement of these parts in the outer surface of a calf
skin, but only includes a small part of the entire
thickness of the corium or fibrous part.

   The Amœba and the blood-corpuscle, living a
wandering life in liquids which furnish their nutrition,
are naked specks of jelly, but many cells coat
themselves with an insoluble and protective layer,
which in animals is usually of keratin or horn substance, and in plants is cellulose. The epidermis
cells are of the keratin-coated kind, and as the
epidermis has no blood-vessels, the cells derive their
nourishment from the lymph and nutritive liquids
of the corium on which they rest. Hence the first
layer is plump and well nourished, and multiplies
rapidly by division, and since the space is limited,
the younger cells push the older away from the
source of supplies, so that they gradually dry up,
lose vitality, and suffer a fatty degeneration, becoming
more transparent, and finally dead. These cells form
the more or less horny surface of the skin, from
which they are removed by friction (and washing),

or, in the parts which are protected by hair, as scales
of scurf, which in moderate quantity are a perfectly
normal product of healthy skin. In parts of the
human body exposed specially to wear, such as the sole
of the foot, or the ball of the thumb, the epidermis is
of considerable thickness, and has been divided by
anatomists into three layers, the living growing part,
sometimes called the "mucous layer," the oily
transparent part, and the "horny" or outer layer;
but from what has been said, it will be understood
that these are not really distinct, but merely stages
in the life-history of the cell. In hairy animals, the
epidermis is extremely thin, probably rarely exceeding $\frac{1}{100}$ of an inch, and it is completely removed before
tanning, the corium or fibrous skin alone going to
form leather.

The hairs shown in Fig. 1 appear deeply rooted
in the true skin, but in reality are completely
surrounded by a layer of epidermis known as the
"hair-sheath," and thus have considerable resemblance to an onion planted on a cloth which is
pushed down into the soil round its bulb. Instead
however of roots penetrating the corium as those
of the onion do the soil, a small knob of the corium
furnished with blood-vessels, and called the hair-
papilla, passes into the centre of the bulb. When
this dies, the hair dries up, its bulb shrinks and it
drops out, but usually a new hair-papilla forms below

the old sheath, from which a young hair originates. In many animals, this process is a regular seasonal change, and the young hairs are a source of trouble to the leather-manufacturer, since their slender pointed tips offer no hold to the unhairing knife, while their vigorous bulbs are more deeply rooted than the old ones. The hair consists of several layers of cells, the outer being overlapping scales like the slates on a roof, while those within are slender and spindle-shaped, and the centre is often filled with a pith of larger and thinner cells, but is never tubular. It is therefore difficult to believe stories of "turning white in a single night," as the hair, once it leaves the bulb, is dead and without circulation, and the dark pigment is very permanent.

Of some importance to the tanner are the sebaceous glands, which secrete fatty matter to lubricate the hair, and which are composed of groups of cells like miniature bunches of grapes situated round the upper part of the hair-sheath, into which they discharge. In the sheep this wool-fat is particularly abundant, and lanoline is a mixture of the purified substance with water. The fat is of a peculiar character, and quite different to the body-fat of the animal, being chemically rather a liquid wax than a true fat. The production of the fat is analogous to that of the degenerating epidermis cells mentioned on p. 12; and its expulsion is assisted by the action of small muscles (*erectores pili*) which also serve to erect the hair; these are controlled by the sympathetic nervous system and are brought into action by cold, fright, and anger. In the human body they are the cause of "goose-skin."

The sudoriferous or sweat glands are also epidermis structures, but are more deeply seated, being usually among or below the hair roots, and discharge through slender ducts lined with epidermis-cells direct to the surface. Though somewhat unimportant to the tanner, they are essential to the life of the animal, and cases are known where death has resulted when they have been stopped by varnish or gilding.

The boundary between the epidermis and the corium is formed by the "glassy" or hyaline layer; a membrane so tenuous that it is almost impossible of microscopic demonstration, and of which the very existence has been disputed. It is at most a mere glaze or varnish to the corium, but the fact that the uninjured surface takes a colour in tanning different to that of portions even slightly abraded, makes it important to the tanner and seems to place its real existence beyond dispute.

In passing from the epidermis to the corium we enter upon a totally different class of structure. Like all animal tissues the whole has been built by cells, but the fibres of which it is composed are cell-products, and not the living cells themselves, which form an inconspicuous part of the mass, being flattened and lying against the side of the fibres. These fibres of "white connective tissue" are identical with those which form the soft flexible framework of the body, and bind together the different organs; and chemically differ but little from the gelatinous constituent of bones and cartilages. Like these, they are converted into gelatine by boiling with water, and only differ from it in their degree of hydration.

Anatomically the greater part of the hide is a felted mass of fibre-bundles of this material, made up of fine constituent fibrils held together by a "cementing substance" differing but slightly from the fibres themselves.

Towards the surface of the skin these fibres are separated into their individual fibrils, which are felted into a compacter and more uniform mass, like the felted surface of a broadcloth.

CHAPTER V

THE CHEMISTRY OF THE SKIN

To understand leather manufacture, something must be known of the chemical as well as the anatomical structure of the skin; and as it cannot be assumed that all my readers are already chemists,

I must be pardoned if I begin with the very elements
of chemical theory.

To the chemist, all the materials of which the
world is composed are built up of atoms of comparatively few kinds, and indeed so far as organic structures
are concerned, we need hardly consider more than carbon, oxygen, hydrogen, and nitrogen, and from these
four alone an infinite number of very different substances, nourishing or poisonous, can be built. Though
an atom is far too small to be seen by the bodily
eye, it is to the chemist a very real thing, and just as
he supposes a diamond to be built up of millions
of similar carbon *atoms*, so he imagines the minutest
crystal of sugar to consist of millions of *molecules*,
each of which is in itself a definite identical structure
of so many atoms of carbon, oxygen and hydrogen.
He knows further that by simple chemical treatment
he can, for instance, so rearrange the structure of
these molecules as to convert the sweet and whole-
some sugar into sour and poisonous oxalic acid, which
contains precisely the same elements. Many instances
are known where two very different bodies contain
not only the same proportions of the same elements,
but even actually the same number of atoms of each.
If then two bodies are constituted of actually the
same atoms, and yet differ radically in their nature
and properties, the difference can only arise from
their difference of arrangement, just as different words

may be formed of the same letters. Efforts to ascertain
the order and connection of atoms in a given substance have to a great extent been successful. Little
is known of the forces by which the atoms are held
together in stable order, and which are often spoken
of as "links" or "bonds"; but which are no doubt
attractions, which if not identical with, are similar to
those of electrified bodies or magnets. It is ascer-
tained that one atom of carbon may satisfy four of
hydrogen or two of oxygen, while nitrogen has a value
equal to three or five of hydrogen. The knowledge
of these satisfying powers, called "valencies," is an
essential step in the discovery of the structure of molecules. In writing such a "structural formula," the atom
of each element is represented by its initial letter, and
the bonds are indicated by hyphens or lines. Thus
marsh-gas in which one atom of carbon is combined

$$\begin{array}{c} H \\ | \\ \text{with four of hydrogen is represented as } H-C-H, \text{ or} \\ | \\ H \end{array}$$

shortly as CH_4. It is obvious that such a formula
does not attempt to indicate the real position of the
atoms with regard to each other, but merely the
order in which they are connected. Even the problem
of the actual arrangement of atoms in space has been
partially solved, but does not at present concern us.
By chemical means it is possible to exchange one or

more of the atoms for other atoms, or other groups of
atoms, and so form new substances of different properties (as for instance chloroform from marsh-gas by
substituting chlorine for three of the H atoms), and
since the new group takes the same position as the
atom it replaces, much can be inferred as to the
relative arrangement of the new compound.   If, in

marsh-gas, a $-\overset{\displaystyle H}{\underset{\displaystyle H}{\overset{|}{\underset{|}{C}}}}-$H group be substituted for one of

the hydrogen atoms, a second hydrocarbon gas is
produced, denser and heavier than marsh-gas, but
otherwise differing little from it in properties.   This
substitution can be repeated again and again, leading
to a chain of carbon atoms, and producing at first
heavier gases, then liquids such as petrol and finally
solid paraffin "waxes."   On the other hand certain
substituted groups produce very definite changes of
properties.   The group $-CO.OH$ substituted for H
in any part of the chain (but usually at one end) at
once converts the compound into an acid, giving the
quality of sourness and the power of combining with
alkalies.   Similarly a substitution of $-O-H$ for H
along the chain gives us sweetness,—glycerine or a
sugar, according to the number of links, and so on.
Again substituting NH_2- (the number of atoms is
conveniently indicated by the small figure) for one or

more of the H atoms of the carbon chain, we produce
what is known as an "amine," a sort of complicated
ammonia which can act as an alkali or "base" and
can combine with acids to form salts.

Now it is possible to substitute one H of a chain
by the $CO.OH$ group, and another by NH_2; and we
then have a compound which is at the same time an
acid and an alkali, though naturally very feeble in
either property. A simple instance is that of amino-
acetic acid $CH_2NH_2.CO.OH$; one of the products
of the decomposition of gelatine and hide (ordinary
acetic acid is $CH_3CO.OH$). Such a double-purpose
molecule has the power of uniting with others of
a like type, the acid head of the one catching the
alkaline tail of the next, and so forming a molecular
chain which may become very long and complex.

Fischer's researches have shown that the proteids,
such as gelatine and white of egg, which form all
animal tissues, including the skin, consist of such
chains of amino-acids, often very complex; and he
has even succeeded in forming some simple proteid
bodies by direct combination in his laboratory, so
that the artificial production of foods is not impossible.

Two principal consequences result from the constitution we have just described, both of which are
of constantly recurring importance in leather manufacture. The first is that the hide-substance, like its

constituents, is "amphoteric,"—that is capable of acting either as a weak acid or a weak base, and consequently of combining with both acids and alkalies, or even with both at once :—the second, that owing to its very large and complicated molecule, it is not capable of forming a true solution like sugar or salt but behaves as a "colloid jelly," of which more hereafter.

Although the foregoing remarks apply to practically all the constituents of hide, it must not be supposed that they are identical. There are broad differences of character between the keratins (horny bodies) which form the epidermis, and the collagen or gelatine-producing substance which constitutes the fibre of the skin, and it is hardly probable that either consists of one individual substance only. Speaking broadly, the keratins are closely related both in constitution and properties to coagulated albumen, of which the commonest type is hard-boiled white of egg. Like this they are insoluble in boiling water, and although at high temperature and under pressure they form viscid solutions, they do not set to jellies on cooling as gelatine does. Keratins are comparatively insoluble in dilute acids, but more easily so in dilute alkalies, forming viscid solutions which are precipitated in curdy flocks by acids or tannins.

The white fibres of the true skin are rapidly converted into gelatine and dissolved by boiling with

water, and form one of the most important raw
materials of gelatine manufacture. They swell, but
do not dissolve in cold dilute acids or alkalies, and
when these are removed by washing or neutralisation
they return to their original shape and size.

Heated with strong acids or alkalies, or under the
action of bacteria, both the keratins and the gelatinous fibres are gradually broken up into simpler
forms, among which are large quantities of the various
amino-acids from which the proteids are originally
built up.

CHAPTER VI

THE PRELIMINARY PROCESSES. SOAKING

BEFORE the actual tannage can take place, much
has to be done which is essential to the success of the
final result. If the hide or skin is fresh from the
butcher, or only lightly salted, the first step is to
cleanse it from dirt, blood and salt, by a few hours'
washing in clean water, preferably assisted by a short
treatment in the drum or wash-wheel, which is a
rotating cylinder, something like a water-wheel turned
inside out, and provided internally with floats or
pegs which carry the hides up above the centre of

the wheel and allow them to fall back, while a stream
of cold water is run on them from a tube in the axle
and escapes through holes in the rim.

If the hides have been dried, the preliminary
washing and softening is much more difficult. Hide
which is dried at a temperature near that of boiling
water cannot be softened sufficiently for tannage by
any known means, and though such temperatures are
never reached in practice, the heat of a tropical sun
is often sufficient to make softening extremely difficult. The old-fashioned method was that of long
soaking, a week or ten days being given in water
seldom changed and consequently full of putrefaction-bacteria, the action of which assisted the
softening, though, as we now know, at the cost of
a serious loss of valuable hide-substance.

The action of even putrid soaks was often insufficient for effective softening, and was assisted by
the "stocks" or "hide-mill" : machines in which
the wet hides were subjected to violent pounding or
kneading for about half an hour, leading to further
loss of partially dissolved hide, and seriously aggra-
vating any blistering or other weakness arising from
unsatisfactory cure. At present soaking is much
shortened, and the necessity for putrefaction and
violent mechanical treatment entirely avoided by the
use of very weak acid or alkaline solutions, both of
which have the power of making hide fibres swell

and absorb water. Alkaline solutions (about 1 lb. of caustic soda or 2 lbs. of crystallised sodium sulphide per 100 gallons), are most generally used, but weak acids, such as formic or sulphurous, are still more efficient, and, in spite of some little practical difficulties, it is very probable that they may ultimately supersede the alkalies, as apparently better weights of leather result from their use.

Salted and dry-salted hides are intermediate between the fresh and the dry; the former require three or four days with as many changes of water; and the latter are soaked like dry hides, but for a shorter time and usually without chemicals. The treatment of skins is quite similar to that of hides, except that, because of their thinness, the softening takes place more rapidly.

CHAPTER VII

UNHAIRING

No doubt our savage forefathers preferred their skin-clothing with the hair on, but for most modern purposes its removal is necessary; and with the hair, that of all the epidermal structures, which prevent the ready penetration of the tan and will not of themselves form leather.

To accomplish this, means must be taken to dissolve, or at least soften the epidermal layer, while leaving the true skin intact. It has been pointed out on p. 21 that the keratins, of which the epidermis is composed, are soluble in dilute alkalies, and it is these which are principally used. Before speaking further of modern methods, however, it is necessary to allude to a very primitive process, which is still in use. Although the harder epidermal structures, such as hair, horns, and hoofs, are very resistant to decay the softer cells, and especially the growing layer next the true skin, are easily attacked by bacterial putrefaction. If a fresh hide be kept for a few days in warm weather, the hair will "slip." Bacteria establish themselves in the soft growing layer, which they liquefy, and so loosen the hair. The difficulty is to confine their action, as the putrefaction rapidly spreads to the true skin and is especially apt to damage the hyaline layer, and so destroy the gloss and surface of the leather : a defect known as "weak grain." The method is therefore now only used in England for sheepskins, of which the wool is less injured than by the ordinary process of liming ; and, in the United States for sole-leather made from dry hides, where slight damage to the grain matters less than weight and firmness ; since even if a portion of hide-substance is liquefied, it remains in the hide and contributes to solidity. The process is called "sweating" or "staling," and is now carried out by hanging the hides or skins in a chamber kept carefully at a temperature of 60° to 70° Fah. with little ventilation and moist air. Much ammonia is evolved by the putrefaction and acts as a weak alkali, taking no inconsiderable part in the unhairing : and, in fact, sheepskins can be very satisfactorily de-woolled by the action of gaseous ammonia in too short a time for serious putrefaction to have taken place.

The more usual process, especially for ox and cow hides, is treatment with milk of lime in large pits. Lime is strongly alkaline, but very slightly soluble in water (about 13 parts in 10,000) so that a dangerously concentrated solution cannot be formed, even if solid lime be used in considerable excess, to maintain the strength of the solution. Only about 3 °/₀ of lime on the wet weight of the hides is actually consumed, but 10 °/₀ or more is frequently added without ill effect ; the quantity required depending a good deal on the extent to which the hides are moved, and the consequent evenness of its distribution. A good method in common use for sole-leather, is the employment of two pits ; all goods going first into a lime through which two "packs" have already passed. In two or three days the pack is moved into a second pit which has only been once previously used, and in the meantime the first pit is emptied, and a new lime liquor made in it, into which the goods are finally drawn, the whole process lasting 7—14 days. It is not uncommon, however, to use one pit only, which is merely "mended" with additional lime for two or three successive packs. The number of times which goods are drawn out and returned to the pits ("hauled and set") during the process also varies considerably: in some yards they are hauled daily, while in others they may remain for two or three days undisturbed. It may be stated, however, that the oftener they are hauled, and the more even and rapid the liming, the less excess of lime is required.

The haulage is usually done with sharp iron hooks on ash shafts 7 or 8 feet long. It is not easy to hook the hides properly through the opaque lime liquor, and there is considerable risk of scratching the surface. A better method is to attach two cords or light chains to each hide, with loops or rings at the ends which are passed over a pair of pegs on the sides of the pit. The hides are successively brought to the surface by the cords, and hauled with short hooks passed through the corkscrew ring by which they are attached to the hide. In many Continental and some American tanneries, the hides are suspended instead of being laid flat, and the lime is circulated and kept in suspension either by a mechanical stirrer or by rocking or swinging the hides themselves. This method considerably shortens the time required, saves hide-substance, and is probably a real improvement, but is more costly in space and first cost, and in the (very small) mechanical power consumed. The liming of sheep and calf skins does not differ in principle from that of hides, but suspension is not usual, and the hauling is done with long-handled iron tongs to avoid scratching.

Although the main object of liming is to loosen the hair and epidermis, it has other important uses. The swelling which it causes separates the fibre bundles into their constituent fibrils, and so increases the fullness and pliability of the leather, which otherwise would be poor and coarse-fibred. Beside simply swelling and splitting up the fibres, the liming removes more or less of the cementing substance between them, so rendering the leather softer and more porous. For sole-leathers this is not desired, and hence the liming is arranged to produce the necessary swelling with as little solution as possible, but for the lighter and more supple kinds, such as are used for the upper parts of boots, more or less removal of the cement-substance is necessary, and must be attained either in the liming or by a later process. To understand how this is possible, the chemistry of the process must be considered.

A new lime is simply a very dilute solution of a strong caustic alkali, and such solutions, when cold, have a powerful swelling action on the fibres, but little solvent action on the cementing substance, or

even the epidermis, and are almost sterile of bacterial
life. By use, however, organic matter is dissolved
from the hides, the proteins of which are split up into
simpler compounds, and the solution then becomes a
nutritive medium for many species of bacteria which
can support its alkalinity.

The action of bacteria is not by any direct attack,
but by the secretion of digestive substances (enzymes),
which liquefy the proteids, so that they can be absorbed by the bacteria, and these digestive enzymes
remain in the lime liquor, increasing its solvent
power upon the hide, while at the same time they
lessen its tendency to swell, in some way not fully
explained. The effect of the ammonia which gradually
accumulates is of a very similar nature, so that the
old lime is a much more powerful solvent than the new,
though it swells less, in spite of its alkalinity being
maintained by the ammonia in addition to its excess
of undissolved lime. The tanner thus has it in his
hands by more or less prolonged treatment in older
limes to remove as much of the 'cementing substance' as he desires. Absolutely new and sterile
limes unhair slowly if at all, so that some treatment
in used limes is essential even for sole-leather, while
for the softer sorts of dressing leather the liming may
be prolonged for a month and takes place mainly
in previously used limes. If, however, the limes are
allowed to become too old and ammoniacal, different

and more actively putrid bacteria are developed which are solvent to gelatine and hide-fibre, and dull-surfaced, porous and light-weighing leather results. The effect of old limes is least injurious in the earlier stages, where the epidermis remains undissolved and protects the hide itself from attack. Hence it is always best to begin the process in the oldest and finish in the newest lime.

Lime is not the only agent which can be used in unhairing, though from its cheapness and comparative safety due to its small solubility, it is most in use. Its action is dependent on its alkalinity, and any other strongly alkaline base may be substituted; though this is rarely done except as a means of "sharpening" or intensifying the action of the lime. The addition of potash to limes in the form of wood ashes was so customary as to have given rise to the German technical name for liming—"*Aescherung.*" Soda (in the form of sodium carbonate which is causticised by the lime) is still occasionally added, where, as in the case of sheep skins for splitting, and of hides for sole-leather, a high degree of swelling is desired.

Caustic soda, used alone, has not been found satisfactory, since, though very rapid in its action, it swells too much and does not dissolve the epidermis sufficiently. A process was some years since patented by Payne and Pullman, in which the hides were first swollen in a 1 °/₀ caustic soda solution, and then treated in one of calcium chloride, which, by double decomposition, replaced the soda by lime, as shown in the equation :

$$2\,NaOH + CaCl_2 = 2\,NaCl + Ca\,(OH)_2\,;$$

but a previous treatment in putrid soak liquors was found necessary to produce satisfactory unhairing.

Although the solubility of lime, and consequently the actual strength of a lime solution, lessens with rising temperature, its unhairing action is much quickened, and many attempts have been made on this account to use warmed limes. Unfortunately the solubility of the hide fibre is increased still more rapidly than that of the epidermis, resulting in lessened swelling and greater solution, and the consequent production of a porous and light-weighing leather. At very low temperatures, on the other hand, the action of lime is almost arrested, and a summer temperature of 15°—20° C. (59°—68° Fah.) is found to give the best all-round results.

In one way only (known as the " Buffalo Method ") has heat been used successfully to quicken liming ; a very short liming being given in the cold, and the hides afterwards suspended in water of about blood-heat. In this way hides may be unhaired in a couple of days with little loss of hide-substance, and will produce excellent sole-leather ; but as they are practically unswollen by the liming, the necessary swelling must be produced by acid, during or before tanning.

Of more importance than the use of heat or alkalies is that of alkaline sulphides or sulphydrates. The effect of mixtures of arsenic sulphides with lime, which by reaction yield calcium sulphydrate, has long been known, and in the East has been used for the removal of superfluous hair from the human body; and a curious account is extant, written by an English nobleman, in the 16th century, who was treated in Italy with this poisonous and evil-smelling mixture. Gas-lime, which in the old method of gas purification contained much sulphydrate, and the sulphide residues of the Leblanc Process have been used; and calcium sulphydrate prepared by passing sulphuretted hydrogen into milk of lime was introduced by Böttger, and proved an excellent depilatory, though little employed on account of the cost and trouble of its preparation. It is only since crystallised sodium sulphide ($Na_2S . 9$ aq.) has been introduced into commerce that sulphides other than those of arsenic have been generally adopted in leather manufacture. The peculiarity of their action is that they do not greatly swell the hide itself, but specially attack the harder epidermis products such as hair, which a strong solution of sodium sulphide reduces to pulp in a few minutes. In this way it is often used for unhairing hides of which the hair is too short to be of

commercial value, the hide being simply dipped into
a strong solution, or brushed over on the hair-side
with one thickened with lime, and then folded, hair-
side in, for a few hours: when the hair may be
removed by brushing, or washing in a wash-wheel.
Similarly such a thickened solution is often used for
unwoolling sheepskins. It is painted on the flesh-
side, and, penetrating the skin, merely attacks the
roots of the wool, allowing it to be "pulled," without
injuring the rest of the fibre.

Most commonly the sulphide is used merely to
quicken the action of the ordinary lime liquors, being
added in small quantities to the pit through which
one pack of hides has already passed. If about $\frac{1}{4}$ lb.
per hide is not exceeded, the hair is not damaged
commercially, and both the time of liming and the
consequent loss of valuable hide-substance are much
reduced. Where sodium sulphide is added to limes,
we always have to take into account the powerful
swelling action of the caustic soda, as well as the
direct unhairing effect of the sulphide. By the addition of calcium chloride to sodium sulphide, the
caustic soda may be converted into common salt, and
replaced in the sulphide by lime, and the effects of
the mixture are then very similar to those of an
"arsenic lime."

The actual removal of the hair, unless it is pulped
by sulphydrates so that it can be washed off, is

effected by scraping with a knife on the "beam," or

Fig. 2.   Unhairing knife.

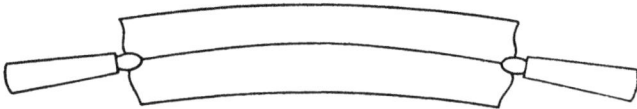

Fig. 3.   Fleshing knife.

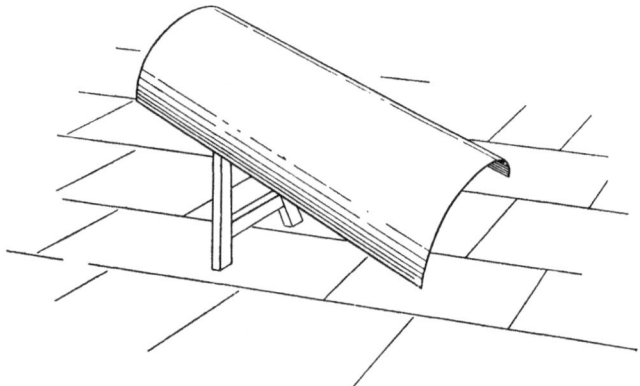

Fig. 4.   Tanners' Beam.

by an equivalent machine. The "beam" is a sort of
steeply sloping table of wood or cast iron of convex

section, over which the hide or skin is thrown, and
the hair is pushed or scraped off by the workman,
leaning over it, and using a somewhat blunt two-
handled knife (Figs. 2, 3 and 4). Machines usually
have a table, or semi-circular drum over which the
hide is thrown, and which takes the place of the
beam, while the knife is a rotating helix. The table
or beam is often padded, and in the very successful
Leidgen machine consists of a stretched apron of
canvas, over which a rotating spiral knife is traversed
by radial arms. An illustration of this machine has
been kindly supplied by the Turner Company of
Frankfort, by whom it is manufactured.

In unhairing it is important not merely to remove
the hair, but, for sole-leather, to work out as much as
possible of the hair root-sheaths and fat-glands, which
are often deeply coloured with the pigment of the
hair. In the lighter and finer leathers this is usually
accomplished by a later operation called "scudding."

The hair being removed, the goods are thrown
into soft cold water, where they swell a little, and
then undergo the further operation of "fleshing";
which is the removal of loose fatty tissue and adhering portions of flesh from the inner or "flesh"
side. This is still mostly done on the same "beam"
as that used for unhairing, but with a double-edged
two-handled knife, the concave edge of which is used
for scraping, while the much sharper convex one is

3—2

Fig. 5. Leidgen's Unhairing Machine.

for cutting.   Machines similar to those for unhairing are also used for this process, but with sharper edges to the spiral knives (Figs. 6 and 7).

So far, the treatment has been similar in principle, if differing slightly in detail, for all classes of leather, from glove skins to the heaviest ox hides, but from this point onward there is greater divergence.   Hides for sole-leather are in England almost invariably "rounded" or, more properly, squared ; since only their middle portion ("butt" or "bends") is strong and solid enough for soles, while the outer parts or "offal" ("bellies" and "shoulders") are more lightly and cheaply tanned and utilised for other purposes. Lighter leathers are also "trimmed," and the irregular "pieces," together with the "fleshings," form a raw material for glue and gelatine manufacture.

CHAPTER VIII

CHEMICAL DELIMING

THE lime has served a useful purpose in loosening the hair and in swelling the hide-fibre, but its removal is necessary before tanning, as it would interfere with that process in various ways, dependent on the particular tannage required.   It is also necessary for the production of soft leathers that the swelling produced

Fig. 6. Fleshing Machine, Front View, Turner Co.

Fig. 7. Fleshing Machine, Back View, Turner Co.

by lime should be very completely reduced, since skin
tanned in a swollen state produces hard leather; and
in some cases a further solution of the interfibrillary
or 'cementing' substance is needed to render the
leather more porous and flexible, and in the extreme
case of glove leathers, so far to loosen the structures,
as to enable the leather to be pulled and stretched
in any direction without a tendency to spring back.
Just as in the case of liming, these various objects
are not attained by special operations, but by modifications of processes, the primary object of all of
which is the complete removal of lime.

As lime and the other alkalies used in unhairing
are all more or less soluble in water, it might be
thought that for this purpose simple washing would
suffice; but since, as has been pointed out in Chap. V,
the constituents of the hide are capable of acting
either as weak bases or acids, they form actual
chemical compounds with the lime, which are only
very slowly decomposed or "hydrolysed" by the
action of water, and the removal of lime by this
means alone would be both tedious and imperfect.
It is true that the old French process for calf skins
trusted only to water and very repeated working
over on the beam with a blunt knife, but the same
water was used repeatedly; this contained bacteria,
and products of putrefaction and hydrolysis of the
skin-substance which themselves acted chemically on

the lime, like some of the fermenting infusions which are used for that specific purpose and will be mentioned later.

In any case, however, a certain amount of loose lime can be, and usually is, removed by washing; but to do this satisfactorily, it is absolutely necessary that the water should be free from "temporary hardness." This hardness, which is present in almost all natural waters, is due to lime or magnesia carbonates dissolved in excess of carbonic acid, and it is these which separate on kettles or boilers as "fur" or "scale," when the excess of carbonic acid is driven off by heat. If a hide containing caustic lime is brought into such a water, the lime immediately combines with the excess of carbonic acid, not only becoming converted itself into insoluble carbonate in the hide, but also precipitating that in the water which had been held dissolved by the carbonic acid. The action in fact is exactly that of Clark's process for softening such waters by adding lime, and recalls the old remedy of "a hair of the dog that bit him," but is expressed to the chemically minded by the following equation :

$$CaCO_3 . H_2CO_3 + CaO = 2\ CaCO_3 + H_2O.$$

The remedy, which is quite effective, consists either in adding a little lime to the water before putting in the hides ; or, what is usually more satisfactory, in allowing perhaps one-third of the lime water from a previous operation to remain in the pit. If this treatment were repeated sufficiently often, the lime would ultimately be washed out; and this, together with the mechanical treatment, is the explanation of the old French method just described. Theoretically, for complete removal an infinite number of changes would be required.

While lime and alkalies are tenaciously retained by the hide, their neutral salts are but loosely held; and suitable treatment with acids capable of forming soluble salts with lime naturally suggests itself as a means of hastening the removal of the latter. The idea is a sound one and largely used, especially in cases where the removal of the lime with as little as possible of any other action on the hide is desired; but it is not so simple in execution as might be supposed.

As the hide-substance is "amphoteric," it is just as ready to combine with acids as with alkalies, and acids equally cause swelling often quite as prejudicial to the tanning process as that of lime itself. It is therefore necessary either to use the exact quantity of acid needed to form the neutral salt, or to take some other means to prevent the skin from absorbing acid in excess.

If "strong" acids are used, such as hydrochloric or sulphuric, it is useless to attempt to get over the difficulty by diluting their solutions, for the hide will absorb a strong acid and swell with it, even from a solution so dilute as hardly to colour litmus red; and the only safe way is to employ an exact quantity of acid barely sufficient to neutralise the whole of the lime, taking, if necessary, other and safer means to get rid of the last traces. Used with sufficient care, quite satisfactory deliming may be attained, sufficient for the preparation of sole-leather hides for tanning, or as a preliminary treatment for the finer leathers which are to undergo a further depleting process; but the safer method is to use acids which are by nature "weak," and which therefore are scarcely capable of swelling the hide.

The terms "strong" and "weak" are here used in a sense which must be unfamiliar to many of my readers, and should therefore be briefly explained. It is easy to understand that an acid like hydro-chloric, which, even if much diluted, tastes intensely sour and violently attacks metals must be "strong," and one like boracic, which does not taste sour, and which even in saturated solution can be used safely to wash an inflamed eye, must be very "weak"; but chemists themselves were long puzzled as to the cause of these peculiarities.

Referring to Chap. V it will be remembered that molecules were described as built up of atoms held together by attractions probably electrical. The

molecule of hydrochloric acid is a very simple one,
consisting of one atom of chlorine united to one atom
of hydrogen. When hydrochloric acid is dissolved in
water, these molecules are to a large extent "dissociated" into their individual atoms, probably by
some sort of association with the water-molecules
themselves, and in this state the atoms can move
freely and independently in the solution, though
they can only be removed from it by neutralisation
of the + and − electrical charges to which their
affinity for each other is due. Such charged atoms
are called "ions." The atomic electrically charged
hydrogen and chlorine in the solution are quite
different in their properties from the neutral and
molecular form in which we know them as gases;
and all those properties of sourness and so on which
are distinctive of acids are due to the dissociated
hydrogen "ion." All acids give rise to such hydrogen "ions" in greater or less proportion according
to their strength; and the greater the dissociation,
the stronger the acidity. The swelling action of
acids on hide, like other strictly acid properties,
appears to depend wholly on the concentration of
the free H-ions, and to be uninfluenced by the varying nature of the negative ions of different acids.
The degree of dissociation differs enormously in different acids. Hydrochloric and sulphuric acids are
almost completely "ionised," even in moderately

dilute solution, while boric acid is hardly appreciably so, as is indicated by its very feeble acidity. In all cases the *proportion* of acid dissociated increases with its dilution, but less rapidly, so that a concentrated solution is more acid than a dilute one, but has also a much larger proportion of undissociated and therefore inactive acid.

Let us consider the neutralisation of an acid by an alkali in the light of these facts. The "strong" alkali, like the "strong" acid, is highly dissociated, but in a different way. Caustic soda, NaOH, for instance, dissociates into + Na and − OH; and it is by the presence of these − OH or "hydroxyl" ions that all the properties we call "alkalinity" are caused; just as "acidity" is caused by the hydrogen-ions. If, therefore, we add hydrochloric acid to a caustic soda solution, we introduce + H ions to a solution already rich in − OH ions, and as H and OH form water, of which the dissociation is almost immeasurably small, the two instantly combine and neutralise, while only Na and Cl remain in the solution, and on evaporation would be separated as salt.

Now suppose, instead of hydrochloric we add boracic acid which is scarcely at all dissociated in solution; its few hydrogen ions immediately combine to form water with the same number of hydroxyl ions. As, however, the dissociated portion always bears a constant proportion to the undissociated in a given solution, it follows that as fast as the H ions are used up, more are produced, and the process goes on, almost instantaneously, to the complete dissociation of the boracic acid and the production of water and of (dissociated) sodium borate. Thus the neutralisation is just as complete and almost as rapid with the weak as with the strong acid, though at no time has there been any appreciable quantity of free H ions in the solution. Even in a strong boracic acid solution the concentration of hydrogen ions is never sufficient to swell hide-fibres, though it neutralises the caustic lime just as effectively as hydrochloric. The advantage of the "weak" acid is therefore self-evident—we can use it even in excess without the least danger; while the undissociated acid acts as a reserve to maintain the strength of the active ions, just as the undissolved lime does in a lime-pit.

Few acids which give soluble lime salts are quite so weak as boracic, but there are many which are so comparatively weak that they can be safely used with much less precaution than the strong mineral acids. The following little table gives the relative "strength" of the acids generally used for deliming under the heading K (a quantity which is known to chemists as the "dissociation coefficient" and which it will be seen varies by some 2000 million times between hydrochloric and boracic acids); and also the

approximate cost at present prices of removing 1 lb.
of lime ; and the quantities of pure (100 °/$_o$) acid
required to dissolve 28 lb. of lime (or in other words,
their "equivalent weights" in lb.).   Wood[1] states
that ordinary wet fleshed hides contain only about
4 lb. of lime per 1000 lb., but the actual quantity of
acid required varies with the liming, the kind of skin,
and the degree of removal which is required.

TABLE I.

Acid	Lb. to neutralise 28 lb. lime	K (100 k)	Cost per 1 lb. CaO
Hydrochloric	36·5	Say 200	1·4 d.
Sulphuric	49·0	,,	0·8
Oxalic	63·0	0·1	8·1
Formic	46·0	0·0214	7·0
Lactic	90·0	0·0138	18·0
Acetic	60·0	0·0018	10·0
Butyric	88·0	0·00115	8·0
Boracic	62·0	0·00000001	6·5

It must be specially noted that while very weak
acids are capable of removing caustic lime, they do
not readily attack any lime carbonate which has
been formed in the skin by the action of hard water
(p. 41), which will therefore remain to interfere with
further operations.

[1] Wood, *Puering, Bating and Drenching*, p. 10.   Spon 1912.

It will be seen that, apart from their disadvantages, the mineral acids (and especially sulphuric) are much cheaper than the weaker organic acids, and the question naturally arises whether with due precaution they cannot be substituted. Alone they are not suitable for complete deliming, since even if used in exact quantities, they act too rapidly and swell the surface of the skin before they have penetrated to its centre. They can, however, be employed indirectly to regenerate the weaker acids with great economy. If, for instance, butyric acid has been used, the liquid will have become charged with calcium butyrate ("butyrate of lime"), and if now sufficient sulphuric acid be added to combine with the lime, the butyric acid will be set free to do its work a second time. Of course, some little knowledge of chemistry is required to do this safely, and a certain amount of butyric acid is carried away each time by the hides and must be replaced, but any intelligent foreman can learn to do the necessary testing and to see that the sulphuric acid is not used in excess; and so long as butyrate of lime is present in the liquor, it is impossible for it also to contain free sulphuric acid; and an incidental advantage is that lime does not beyond a certain point accumulate in the liquor, since the sulphate is very sparingly soluble.

Although the acids named in the preceding

paragraph are all relatively "weak," yet, as shown by
the table, they differ widely in this respect, and even
the weakest except boracic is capable, if carelessly
used, of swelling the skin to a dangerous extent, while
boracic acid, though free from this risk, gives somewhat insoluble lime salts and is often unsuitable for
other reasons, and mainly employed for sole-leather.
Fortunately a means exists of still further weakening
any of the weaker acids to any required extent,
simply by the addition of their own neutral salts.
The explanation of this is that, just as lime, for
instance, only dissolves till a certain concentration,
known as a "saturated solution," is reached, so acids
only produce active ions till these reach a certain
concentration peculiar to the acid, high in strong
acids and low in weak ones, and proportional to
the product of multiplication of the H-ion by the
acid-ion. The salts even of weak acids have high
ionisation-concentrations, yielding the same acid-ion,
but a base-ion instead of the H-ion to which acidity
is due. Hence if a salt is added to the acid-solution
the acid-ion is greatly increased, and as the product
of multiplication of the acid-ion by the H-ion is
constant, the latter is diminished, and with it, the
active acidity of the solution. Thus acetic acid,
which reddens congo-red and swells hide vigorously,
ceases to do either if sufficient sodium acetate is
added, while its power of dissolving lime is not

diminished. As the deliming operation with any acid soon produces considerable quantities of its neutral lime salt, all that is necessary to produce the desired weakening effect is to strengthen the old liquor, or a suitable proportion of it for use on a second pack instead of making an entirely fresh one[1].

Unfortunately this method of weakening is only available for acids naturally weak and has little effect on the strong mineral acids, but it is quite effective when a mineral acid is used as described merely to liberate a weak one from its salt.

Another method of chemical deliming which is effective, and a good deal used, depends on the employment of the salt of a "strong" acid in combination with a "weak" base; of which ordinary sal-ammoniac (ammonium chloride) is a typical example. When a limy hide is brought into a very weak solution of this salt, the lime immediately combines with the chlorine, setting free ammonia, which, as a much weaker alkali than lime, swells the hide less and is comparatively readily washed out. This effect in a slightly different form occurs naturally in the old-fashioned fermentive "puers" and "bates."

[1] The subject is fully discussed by Stiasny in an article *Anwendungen des Massenwirkungsgesetzes auf einige gerbereitechnische Vorgänge*, Collegium 1912, p. 289, and also in English in *Journ. American Leather Chemists' Assoc.* 1912, p. 301.

The effect of removing lime by any of the chemical means which have been described, is at once to reduce the swollen and tense condition of the limed hide to one of comparative softness, sufficient for sole and belting and harness leather, but not so for the finer leathers. The removal of lime allows the fibre to return only to the natural slightly swollen condition of gelatinous matter saturated with cold water. With tepid water this swelling is somewhat further reduced, but still not sufficiently for the finer leathers. When the hide "falls," or loses its swelling, the gelatinous fibres part with water, but the hide as a whole does not do so to the same extent, for as the fibres contract the interstices between them increase and become filled with the liquid. The cardinal difference between swollen and unswollen hide is that in the former the water is *in* the fibres in jelly form, while in the latter a much larger proportion exists as liquid between them, and can be squeezed out by very moderate pressure.

In order still further to reduce the swelling of the hide fibres, the tanner has recourse either to fermenting infusions of bran or of animal excrements, or to artificial products which yield a similar result. The cause of their effect is by no means clear, but before discussing it we must say something on the general subject of fermentation, of which the putrefaction of animal matters is merely a special case.

CHAPTER IX

BACTERIA AND FERMENTATION

THE principal agents in the fermentations important to the leather manufacturer are bacteria, though, as regards sugars, yeasts play an important part. Although bacteria are matters of everyday talk, yet the popular conception is frequently so vague that some description of their character and life-history must be given.

Bacteria are single-celled vegetable organisms, so small as to approach the limit of microscopical visibility, or even to fall below it: so that little is known of their internal structure. They vary in shape from minute spheres to slender rods sometimes curved or spiral, and frequently of a dumb-bell form, thinner in the middle than at the ends. They multiply by lengthening and division, the spheres becoming elongated before dividing; and not unfrequently the divided cells remain adhering in the form of chains. As division may occur so frequently as once in 20 minutes, their multiplication is extremely rapid. Some few also, under conditions unfavourable for growth, can form single spores or reproductive cells, which are much more resistant

than the bacteria themselves, and may retain vitality
and the power of development for years, after drying,
freezing, or even short exposure to a boiling temperature; and are also extremely difficult to destroy
by disinfectants. Such spores are the most dangerous feature of the anthrax bacterium which causes
the dreaded "malignant pustule" and "woolsorters'
disease." Many bacteria swim actively in liquids, by
the aid of slender threads of protoplasm (*flagella*)
protruded from the cell.

All bacteria feed on and alter the nourishing
liquids which surround them, and each requires its
own particular kind of food, and produces in it its
own especial change. Of the many thousand species
known, only very few are actively harmful to man-
kind, either by producing poisons in the human body
or destroying its tissues ; and a large number are
positively beneficent, if not essential, to animal life;
so that the popular idea that all must be destroyed
is a false one.

Bacteria have no mouths and must therefore
always have liquid nourishment which they can absorb, and those which attack solids, such as the skin,
do so by secreting solvent or digestive liquids known
as *enzymes*, or "unorganised ferments" (in distinction
from the organised bacteria themselves). Such ferments are not peculiar to bacteria, but are made
useful in many phases of both animal and vegetable

life. The active agents of animal digestion, pepsin, trypsin and the rest, are of this nature, and the plant uses them to dissolve, and render available for nourishment, the starch and other foods stored in the tuber or the seed. Enzymes are purely chemical substances without life, and can be separated by precipitation and preserved for years in an active state; but a surprising peculiarity is that they are not altered or destroyed by their action, but in limited quantity can cause an indefinite amount of special chemical change. They are, in fact, "organic catalysts." They cannot build, as living organisms often can, but can only pull down; and the changes they produce are those which tend to take place, and probably do very slowly take place, without them. It is possible that their action, like that of many inorganic catalysts, is simply that of facilitating the interchange of the electric charges by which molecules are held together. Their usual mode of action is that of breaking up by hydrolysing, or adding water; and is well illustrated by the invertase of common yeast, which breaks up cane sugar into the simpler glucose, on which alone the yeast can feed.

$$\underset{\text{Cane Sugar}}{C_{12}H_{22}O_{11}} \quad + \quad \underset{\text{Water}}{H_2O} \quad = \quad \underset{\text{Glucose}}{2\ C_6H_{12}O_6}$$

The changes which take place *within* the cells of the bacteria are usually of much more complex character, and though the main products are of

simpler forms than the nutriment, very complicated
substances may also arise, such as organic poisons,
colouring matters, enzymes, and the protoplasm of
the bacterial cells. As with animals, the excreted
products are poisonous to the bacteria themselves,
and limit or prevent their increase. Many bacterial
fermentations are very complex; they may be begun
by one species, which dies, poisoned with its own
products, or by having exhausted its nourishment;
the process is then carried on by others to which
the first gives place, and which can consume as food
the products already formed, breaking them down to
still simpler forms.

Bacteria are not the only organisms which cause
fermentations. Yeasts, which are larger globular or
elongated cells, some moulds, and even some simple
forms usually considered animal also take part in
such changes, but (except as regards form) most of
what has been said about bacteria is true also of
these, and need not be repeated.

CHAPTER X

THE FERMENTIVE "BATES"

In the earlier times, washing and working with
more or less stale waters, as described on p. 23, was
probably the only means used to remove lime, and

bring down the skin to the "fallen" and flaccid state
which is necessary for the production of soft leathers.
The various fermentive processes were apparently
not known in England before the 18th century; or
if known, only as secrets; and the earliest account
of them is in a book entitled *The Art of Tanning
and Currying Leather*, published in London in 1780.
Apparently, like many of the older processes of
leather manufacture, the methods were derived from
the East. In the present day, treatment with fermenting bran infusion is called "drenching," that
with pigeon or hen dung "bating" (possibly abating),
and that with dog dung "puering" (no doubt from
the Fr. *puer*, to stink). In earlier times all these
infusions were called "masterings," and various other
fermentable matters were used beside those named[1].

Although, in practice, the bran drench is mostly
used to cleanse the puered skins, to bring them into
a slightly acid state, and to complete the deliming
before skins go into the tanning liquor, and rarely as
an independent method, it is probably the simplest
of the fermentive processes, and as an acid deliming
method is closely related to those which have been
already discussed, and on these grounds may be first
described.

The fermentation in the bran infusion is not of
proteids but of carbohydrates (sugars and starches),

[1] The modern Fr. term is "*confit*," the Ger. "*Beize*."

and its important products are acids, mainly lactic
and acetic.

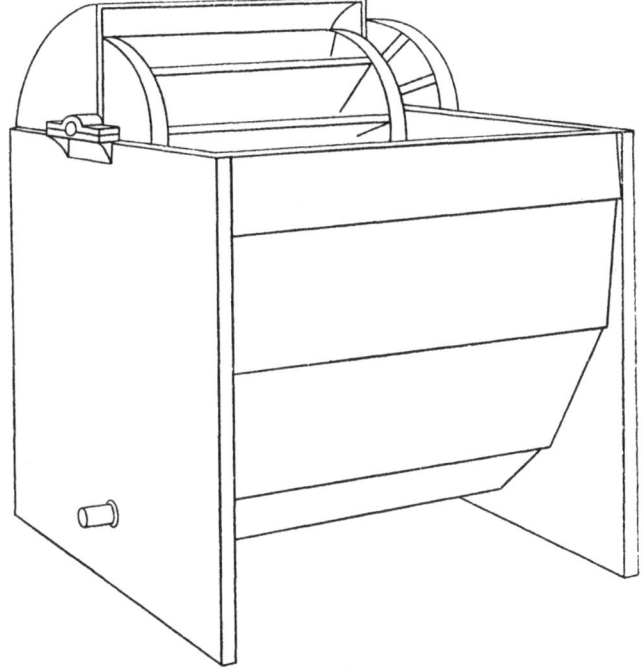

Fig. 8.   Paddle-vat.

The process consists in bringing the skins into an
infusion of bran, and allowing them to remain there

during the bacterial fermentation : this has been very fully studied by J. T. Wood, to whose book[1] the reader may be referred.

Five to 10 °/$_o$ of bran on the weight of pelt, and $\frac{1}{2}$ to 1 °/$_o$ on that of the water is generally used, though practice is somewhat variable. In England the bran is usually mashed in water at a temperature of 95° Fah., and the skins are put in and paddled either in a tub by hand or in a vat such as is shown in Fig. 8 till the temperature falls to 85°—83° Fah. and the fermentation begins. No addition of ferment is necessary, as sufficient is always present in the vats. A good deal of gas (principally hydrogen with some nitrogen and carbon dioxide) is produced by the fermentation, both among and actually in the skins, so that the latter float up at intervals, and have to be put down with a stick or the paddle-wheel. This effect is known as the "working" of the drench and is a rough indication of the amount of fermentation, so that the skins are known to be sufficiently drenched when they have risen two or three times. If the drenching is continued too long, small blisters are produced by the evolution of gas actually in the texture of the skin, which ultimately burst, and produce "pinholes." The skins, on coming from the drench, are soft but somewhat full, containing much water between the fibres, and milky

[1] *Puering, Bating and Drenching of Skins.* Spon 1912.

white. To the expert, feel and appearance are the surest guides to correct drenching.

Wood's explanation of the chemistry of the drench is that the starch is first converted into glucose and other fermentable sugars by an enzyme naturally present in the bran, and that these sugars are fermented by the bacteria, producing lactic and acetic acids, which remove the lime. The process, when proceeding properly, is to a large extent self-regulating, since the bacteria are very sensitive to an excess of acid, and when a certain concentration is reached (2 or 3 grm. of mixed acids per litre) the production is automatically checked till a part of that formed is absorbed by the skins. The fermentation is mainly due to two species of lactic bacteria which Wood has named *Bacterium furfuris*, a and β[1].

These bacteria feed only upon carbohydrates, and are incapable of attacking the skin, so that if they alone were present the operation would be a safe one. It is, however, obviously impossible, under practical conditions, to secure this, and a working drench always contains many other species, some of them harmful. To prevent these seriously attacking the skin, the only way is to make the conditions so

---

[1] These are dumb-bell-shaped organisms, not exceeding $0.75\,\mu$ long, and from 0.1 to $0.5\,\mu$ broad. 1μ or $\frac{1}{1000}$ of a millimetre is about $\frac{1}{25000}$ of an inch, so that a chain of these organisms one inch long would contain nearly 40,000.

favourable for the proper bacteria that they are able to overgrow and keep down the others. Most bacteria which attack skin only thrive in alkaline solutions, so that the faint acidity of the drench is a great protection, and many harmful organisms multiply most rapidly at higher temperatures, so that it is important to keep all conditions of temperature, acidity and food as favourable as possible for the drench bacteria; to renew the drenches for each pack; and to be sure that sufficient of the right bacteria are present to start at the outset a vigorous fermentation. In spite of all precautions, the process is not without risk, and under unfavourable weather-conditions either very acid or putrid fermentations occasionally set in, which destroy or damage the skins. For this reason it is probable that drenching will ultimately be completely superseded by chemical deliming, such as has been earlier described.

Both the chemistry and the bacteriology of "puers" and "bates" are much more complicated than those of the drench; the bacteria are more varied, and the proteids fermented more complex; and it is possible that not only enzymes produced by the bacteria but those originally present in the dung are involved. Although the principles of action of both dog and fowl dung are almost identical, their practical use and effects materially differ. The

dog-dung puer is always used warm, as its efficient
bacteria require a temperature approaching blood
heat for their free development; and its effects are
very rapid. Bates, on the other hand, are usually
employed cold, though their action is increased by
warming; and their effect is less violent than that
of puers, and much more gradually produced. They
are therefore suited to hides and thick skins, which
are penetrated evenly and completely, though less
thoroughly "pulled down" than thin skins are by
puering. If puering were applied to such heavy
goods, the surface would be damaged or even dissolved before the action had time to reach the
centre.

The dung is largely obtained from kennels, but
is also collected in the streets, and dry dung from
the East is an article of commerce. Before use it is
made into a soft paste with water, and allowed to
ferment for some weeks; and it will retain its activity
in the paste condition for months, though when
exposed to the air in a solid but moist condition,
it rapidly deteriorates and becomes useless from
putrefactive changes. The paste, when in good condition, is of a pale buff colour, and in use enough is
added to water at a temperature of 110° to 120° Fah.
to produce a turbid but not soupy liquid, about
10 °/₀ of the paste being required on the weight of
the wet skins. The liquid is allowed to cool to 100°

Fah. or lower before putting in the skins, which are often previously warmed in tepid water to avoid chilling the liquor. During puering the skins must be frequently moved to ensure even action, and for this reason a paddle-vat (p. 57) is usually employed. The process is usually complete in 1—2 hours at most. The skins, which should be (even if partially delimed) in a slightly alkaline condition, very rapidly lose their firmness and become so soft that they fall in folds in any direction, and the common statement that a puered lambskin can be drawn through a wedding ring is scarcely an exaggeration. It is hardly possible to give any criteria for the completion of the process, which is recognised by the practised eye and touch, and differs for different kinds of skins ; but if allowed to go too far, the skin is injured, and ultimately entirely dissolved or digested. Apart from this avoidable risk, damages are very liable to occur from the presence of wrong bacteria, or from allowing the skins to be too long unstirred, in which case the action is irregular, and stains and marks arise from the settling in the folds of mud containing pigment-producing or gelatine-liquefying bacteria. The desired action is not merely a deliming but a digesting one, designed to remove a portion of the inter-fibrillar substance, and to soften and loosen the texture ; and it is obvious that the operation, disgusting as it is, involves considerable

skill and experience. The warm liquid has a sickly
and very unpleasant smell, which clings obstinately
to the hands, the clothes, and even the hair of the
operator, though it cannot be said to be unhealthy.

The composition of dung even when "pure" and
fresh is a very mixed one, containing unused digestive ferments and bile products from the animal,
bacteria which exist in variety in the bowels and
actually take part in digestion, and masses of partially
decomposed proteids and carbohydrates which are un-
fit for animal nutrition, including amino-acids, amines,
and amine salts, and, especially in dogs, large quantities
of phosphates of lime and other bases. Additional
bacteria of many kinds enter it from the air and
from the vessels containing it, and during the period
of preliminary fermentation, produce fresh enzymes
and carry the process of digestion further, bringing
many of its solid constituents into a soluble form.
Wood mentions over 90 different species of bacteria
which have been isolated; but apparently the species
most active in the puering process is that isolated
by Becker and called by him *Bacillus erodiens*,
probably a variety of *B. coli*, a very common and
harmless intestinal bacterium, but so much resembling
the typhoid bacterium that it has frequently been
mistaken for it. *B. erodiens* is very similar in shape
and appearance to the drench bacterium, and is
capable of pathogenic action if introduced into the

blood, so that caution should be taken not to allow it to gain access to wounds, though the author has never heard of any serious results. Though probably the most active in puering, it is not the only organism involved, and as it does not produce enzymes capable of attacking the skin, its action must probably be supplemented by other species. It is mainly anaerobic, does not liquefy gelatine, and, like the drench-bacteria, produces gas, partially hydrogen in the fermentation, and is capable of fermenting glucose, and producing acids.

The chemical constituents of the puer-liquor are themselves capable of some depleting action, even in the absence of living bacteria and of enzymes, both of which may be removed by boiling and filtering. Wood found that the amine salts were the principal agents in this action, and produced a similar result with a pure solution of amine chlorides. Amines are merely more complex and less alkaline ammonias, and their effect is therefore similar in principle to that of ammonium chloride (sal ammoniac) which has been described in a previous chapter. The soluble organic phosphates also take part in the process, forming insoluble calcium phosphate with the lime present in the hide.

To produce the full puering effect, some solution of the hide substance is demanded, and this is accomplished by the enzymes. The enzymes

cannot be separated in a pure state, but are precipitated together with other albuminoids if the
liquid containing them is added to a large volume
of absolute alcohol; and on redissolving in water,
the precipitate is found to be still active. In this
way Wood separated the enzymes from a filtered
puer liquor and found that a solution free from
bacteria, but containing $\frac{1}{2}$°/₀ of the enzymes and $\frac{1}{2}$°/₀
of amine chlorides, produced an effect quite similar
to that of the puer, though either constituent alone
had a much less effect. In the case named, both
enzymes and amines were prepared from the puer
itself, but quite similar effects were obtained by
those from bacterial cultures which contained no
dung.

The last fact was not without interest in view of
a suggestion made by W. J. Salomon in 1892, that
the effect of puer was due to digestive enzymes
derived from the animal, and especially to pepsin
and the pancreatic ferments. The continued existence of digestive ferments after the dung has
undergone fermentation seems unlikely, since they
are very putrescible, but recent attempts have been
made, in view of patent litigation, to prove their
presence. Wood found (*Jour. Soc. Ch. Ind.* 1894,
p. 218), as was to be expected, that pepsin was quite
inoperative, since it is only active in acid solution
but that "pancreatin" (a mixture of the various

pancreatic ferments) had a very marked though
incomplete puering action, even when bacterial
activity was prevented by chloroform. Wood's suggestion has since been the subject of a successful
German patent taken by Dr Röhm; but, at the time,
his failure led him to turn his attention from the
pancreatic ferments to the preparation of an artificial
bacteriological puer. Dr Becker of Frankfort had
also been investigating the subject, and as both
arrived at practically identical results, a joint patent
was taken out, and the product has been sold and
much used under the name of "Erodin." It consists
of a suitable nutritive mixture of gelatinous matter,
peptonised by heating with acid, with the addition
of some phosphates. This is dissolved in hot water,
and at the proper temperature a special culture of
bacteria (largely *B. erodiens*) is added and is allowed
to ferment before entering the goods. The result is
perfectly successful for calf skins, and much less
dangerous to the skins than the use of dung, but the
smell is not less objectionable, and on certain classes
of goods the effect appears inferior to that of actual
puer.

The pancreatic bate introduced by Dr Röhm has
also similar limitations. It contains some pancreas
extract with a considerable amount of ammonium
chloride, and acts without bacterial life and its
attendant evil smell and danger to the skins; but

it cannot be said to have finally solved the entire
problem. Much experiment is still going on, and
enzymes other than those of the pancreas are being
used; and there is little doubt but in the near future
the use of dungs will be banished from leather manufacture.

CHAPTER XI

THE CONVERSION OF SKIN INTO LEATHER

ALL the processes so far described have for their
object the removal of the epidermis and its ap-
pendages, and the cleansing and purification of the
corium, in preparation for its actual conversion into
leather. The skin is still merely raw animal matter.
Moist, it is soft and pliable but rapidly putrefies;
while if dried it becomes like horn, stiff and trans-
lucent and useless for most of the purposes of
leather. The problem then is to render it soft, porous
and opaque when dry; imputrescible and sufficiently
resistant to water; and this can be done in a variety
of ways.

The cause of the horny nature of dried skin is
that the gelatinous and swollen fibres of which it is
composed not merely stiffen on drying, but adhere

5—2

to a homogeneous mass, as is evidenced by its trans-
lucence. If in some way we can prevent the adhesion
of the fibres while drying we shall have made a step
in the desired direction, and this will be the more
effective the more perfectly we have split the fibre-
bundles into their fine constituent fibrils, and removed
the substance which cements them. This is largely
accomplished in liming, puering and bating. The
separation of the fibres during drying can be partially
attained by purely mechanical processes. Parchment
and vellum are products of this description, but their
softness is quite insufficient for leather, and, if soaked
in water their fibres swell and again adhere. Knapp,
to whom we owe our first intelligible theories of the
tanning process[1], showed that by physical means the
separation and drying of the fibres could be so far
effected as to produce without any tanning agent a
substance with all the outward characteristics of
leather, although on soaking it returned completely
to the raw hide state. He soaked the prepared
"pelt" (as skin in this wet condition is called) in
absolute alcohol, which penetrated between, and
separated the fibres, and at the same time dried
them by its strong affinity for water. More recently
Meunier has obtained a similar result by the use of
concentrated solution of potassium carbonate which
is even more strongly dehydrating.

[1] *Natur und Wesen der Gerberei und des Leders*, Brunswick, 1853.

Knapp made a further step by adding to his alcohol a small quantity of stearic acid which, as the alcohol evaporated, left a thin fatty covering on the fibres which completely prevented their adhesion, and reduced their tendency to absorb water; and he so produced a very soft and white leather. Somewhat similar are the principles of the many primitive methods which apply fatty and albuminous matters, grease, butter, milk, or brains to the wet skin, and by mechanical kneading and stretching, aided by capillarity, work these matters in between the fibres as the water evaporates. Such methods are still used for laces, thongs, and furs; and enter into many processes in which other tanning agents are also employed.

Building upon these facts, Prof. Knapp advanced the theory that the effect of all tanning processes was not to cause a change in the fibres themselves, chemical or otherwise, but merely to isolate and coat them with water-resisting materials which prevented their subsequent swelling and adhesion. True as this theory undoubtedly in many cases is, it can hardly be accepted as the whole truth, and it seems incontestable that frequently the fibres themselves undergo actual chemical changes which render them insoluble and non-adhesive.

Before Knapp's work, the prevalent theory, at least as regards vegetable tannage, had been a chemical

one, started by Sir Humphrey Davy. If a solution
of gelatine be mixed in proper proportion with one
of tannin, both unite to form a voluminous curdy
precipitate; and, according to Davy's ideas, this was
amorphous leather. Against this, it was urged that
even the supposed "tannate of gelatine" itself could
not be a true chemical compound, since the proportions of its constituents were considerably
varied by changes in the strength of the solutions, or by washing the precipitate with hot
water; and further, that in chemical compounds,
the form was changed, and no trace of the original
constituents appeared in the compound; while in
leather, apart from some change of colour and
properties, the original fibrous structure remained
unaltered.

This reasoning appears much less conclusive now
than it did in Knapp's day. Against the last objection guncotton may be quoted as an instance of
profound chemical change with no alteration in outside appearance; and it is recognised that, especially
among complex organic substances, chemical reactions
are rarely complete, but that stable positions are
reached, so-called "equilibria," in which the proportion of changed and unchanged substance is
dependent on concentration or other conditions; and
that therefore such a precipitate might well be a
mixture of gelatine with a true gelatine tannate

from which further portions of tannin might be dissociated by water.

With the clearing up of old difficulties, however, the conflict between chemical and physical theories has, as is usually the case, merely passed into a new phase. Years ago, it was shown by Linder and Picton and others, that liquids could be obtained which were not really solutions of ions or molecules, but merely suspensions like that of clay in water, or butter-fat in milk ; but so finely divided as to appear clear and transparent, and pass through filters like true solutions. Later, by means of the ultra-microscope their discrete particles have actually been made visible, each of them consisting of many molecules of the suspended substance. Nevertheless, these particles have many molecular properties, possessing + or − electrical charges; behaving like large ions under the influence of an electric current ; and mutually precipitating and neutralising each other when negative and positive are brought together. Such solutions are called " colloid," and those of gelatine and tannin are of the class, so that it is now often said that the precipitation of gelatine by tannin, and the fixation of tannin by gelatinous fibre are merely "colloidal" and "physical," and not "chemical" phenomena. Admitting the facts, the question still arises whether the distinction between chemical and physical is not here one without a

difference; and whether between the purely ionised
dilute solution of a salt and the coarsely granular
clay suspension there is any point where a definite
line of demarcation can be drawn. The writer in-
clines to the view that there is not; and that ionic
and colloidal combinations are extreme cases of the
same laws; both chemical, and both physical. It
will be seen, when the details of tanning processes
are considered, that in some one limit and in some
the other is approached.

CHAPTER XII

THE PICKLING PROCESS

THOUGH the pickling process is strictly one for
preserving the prepared pelt for tanning, rather than
of converting it into leather, it is in itself capable of
producing leather; and its investigation has thrown
so much light on the actual tanning processes, that
a brief notice seems in place before their con-
sideration.

"Pickling" consists in a preliminary swelling with
acid, which is afterwards reduced by a concentrated
solution of common salt; and the skins so treated
can be kept for many months in a wet condition

without putrefaction or injury. It was originally
applied almost exclusively to the preservation of the
split grain-sides of sheepskins, and, later, of the
entire pelts; large quantities of which are now imported from New Zealand in this condition. It
has also acquired considerable importance as a
preparation of these and other skins for chrome
tanning; and is sometimes used as a means of
deliming.

The usual method is to employ as a "rising liquor"
a solution containing about $8°/_o$ of salt with $\frac{3}{4}°/_o$ of
sulphuric acid; the object of the salt being to prevent
excessive swelling. The skins are stirred or paddled
in this bath for half or three quarters of an hour,
and in spite of the salt, swell considerably. They
are then transferred for a similar time to a solution
of common salt kept saturated by excess of solid
salt, and in this they become very thin and white,
and are ready for packing in casks. If a skin in
this condition is allowed to dry, and softened by
stretching and "staking" (drawing over a blunt
edge), it forms a very soft and white leather, quite
permanent so long as it is kept dry. If it is put in
water, however, the salt is removed, and the skin
swells excessively, and rapidly becomes tender and
rotten, and if tanned in the swollen condition the
leather is useless and tears easily. It is therefore
necessary in tanning pickled goods, either to use at

first salted liquors or to neutralise the acid by an
alkali before washing out the salt. A skin which has
been pickled is always softer and more porous after
tannage than an unpickled one, owing to the very
complete differentiation of the fibre-bundles; and
in chrome tannage the process is used for this
object. Other acids than sulphuric can be used, and
especially formic acid in $\frac{1}{4}$°/₀ solution gives an excellent result, while a common German fur-dressing
method depends on lactic acid developed from fermented rye-flour. The author has however recently
shown that whatever acid is used for swelling, that
combined with the skin after salt-treatment is mainly
hydrochloric. He has also devoted much time to the
investigation of the causes of the swelling caused by
acids, and the still more powerful contraction produced by salt. Since in ordinary wet hide the liquid
exists partially in the jelly form in the fibres, and
partially in the interspaces between them, it is impossible to make any accurate determination of the
swelling of the fibres themselves; and for this reason
most of the experiments were made upon purified
gelatine, which chemically is almost identical with,
and which swells and contracts in the same manner
as, hide-fibre, though not perhaps exactly to the same
extent. Pure hydrochloric acid was used in place
of a mixture of sulphuric acid and salt to avoid
complication of a mixture of different acids; and

check-experiments were made with actual hide to
prove that the effect was qualitatively if not quan-
titatively the same. One grm. of pure dry gelatine
was soaked in a given solution till equilibrium was
attained, and no further change took place: for
which 48 hours was found sufficient. By weighing
the swollen jelly, the gain in weight showed the
amount of liquid absorbed and the consequent swelling; and both the liquid and gelatine were analysed
to determine the amount of acid and salt contained
in each.

The first point to determine was the law of swelling
with acids alone. With hydrochloric acid the swelling
increased rapidly at first with increase of concentration of acid, but a maximum was soon reached, and
further concentration produced a steady diminution
of volume, to a point at which the experiments were
discontinued owing to the commencing solution of
the jelly in the acid, which begins at a concentration
of about 0·3 N (0·3 gramme-mols. per litre); while
the maximum swelling was observed at a concentration so low as about 0·006 N. One grm. of the
gelatine used absorbed about 8 grms. of pure water,
over 50 grms. of the 0·006 N acid, and only
18 grms. of 0·2 N. The concentration of acid in
the gelatine was always in excess of that in
the aqueous solution by an amount which, after
the maximum swelling was reached, was almost

constant at 0·8 milligram-mols. of HCl per grm. of
dry gelatine

(i.e. of $36·5 \times 0·8 = 0·0292$ grm. HCl).

The constancy of this amount points very clearly to
the formation of an actual chemical compound of
the nature of a gelatine chloride, rather than to any
merely physical adsorption which would increase
continuously with concentration. It would occupy too
much space to discuss either the experimental work
or the reasoning which led to the conclusion that
an actual hydrolysing and ionising salt was formed
which followed closely Ostwald's hydrolysis formula,
and that the swelling was caused by the osmotic
pressure of the Cl ions of this salt, and repressed
by, that of the Cl ions of the acid outside the jelly.
For detailed discussion the reader must be referred
to the original papers[1].

Other strong acids behaved similarly to hydrochloric, showing a very marked maximum, which
with weaker acids diminished and finally disap-
peared. Thus formic acid showed a distinct maximum,
while with acetic acid the swelling continuously
increased with the concentration of the acid. This
behaviour is in complete accordance with the theory

[1] "Ueber die Einwirkung verdünnter Säuren und Salzlösungen
auf Gelatine," *Kolloidchemische Beihefte*, Bd. II, Heft 6, 7, p. 243;
and *Journ. of American Leather Chemists' Association*, p. 270, v,
1911; *Trans. Chem. Soc.* p. 313, vol. 105, 1914.

which has been suggested; a greater concentration of a weak acid being needed to produce a stable salt, while the acid-ion cannot reach sufficient concentration to produce marked repression of the ionisation of the gelatine-salt. While it is thus implied that the swelling action of acids is governed only by their ionisation, it does not follow that for the purposes of the tanner it is indifferent what acid is used. Not only are the weaker acids easier and safer to handle, but it has been shown that in actual tannage the acids are wholly or mainly replaced by the tanning substances, and this replacement must take place the easier the weaker the acid, and, conquently, the more readily its salt is hydrolysed or decomposed by water.

It is also possible that the salts of hide substance with different acids have different characteristics, but as they are mainly, if not entirely, decomposed in the further stages of tanning, it is probable that acid "strength" or "weakness" has a greatly preponderant effect.

A fact however which at first sight seems puzzling, is that common salt will produce powerful contraction of skin or gelatine swollen with *any* acid, even a weak organic one. Perfect pickling can be obtained with salt and formic, or even acetic acid; though the concentration of the acid required is somewhat greater than with the stronger acids. This difficulty disappears on reflection. In a dilute solution containing salt and hydrochloric acid, in which both may be taken as completely ionised, we have a mixture containing Cl-ions, H-ions, and Na-ions, all independent, except that the sum of the negative ions must equal that of the positive. If for sodium chloride we substitute sodium formate, HCO_2Na, we have a new acid ion HCO_2 in addition to those originally present, but equally free. If we increase the concentration till un-ionised acids and salts are formed, these latter will bear a fixed relation to the quantities of their respective ions present in the solution, no matter whether these were introduced as acids or salts. Thus, similarly, if we bring gelatine formate into a concentrated solution of sodium chloride, rearrangement will take place simply according to the concentration of the ions present, and if the salt is present in large excess we shall have much gelatine chloride and little formate in the outer solution; thus forming what is called a "quadruple equilibrium" in which each salt of gelatine in the skin is balanced and compressed by its own salt of sodium in the solution.

It has been proved by actual analysis, that skin swollen with formic acid, and reduced by salt, contains practically as much hydrochloric acid as if this acid had actually been used in swelling.

The action of the salt solution in pickling is not,

as has been supposed, to remove acid from the skin,
unless the acid has been used in excess; but by
repressing the ionisation and consequent hydrolysis
of the "skin-chloride" it actually enables more to
be fixed; a point which is of importance in explaining
the *rôle* of salt in alumed leathers.

It is obvious that much which has been said of
the acid swelling of gelatinous matter must be
equally applicable to alkaline swelling such as that
of the limes, and to the contraction produced by
deliming and bating agents.   There is no doubt that
chemical compounds are produced in this case also,
but with the difference that the protein acts as an
acid instead of a base, and that its carboxyl- and not
its amino-groups are active, and the OH' ion takes
the place of H.   A marked difference is that sodium
chloride seems to have no compressing effect, even
when the swelling is produced by sodium hydrate,
proving that the metallic base takes no leading part
in the swelling which however is repressed, after
reaching a maximum, by increasing concentration of
OH'.   The writer hopes to take up the study of
alkaline swelling as soon as his investigation of acid
swelling is completed.

CHAPTER XIII

ALUMED LEATHER

ALTHOUGH, in Europe at least, the production of alumed leathers is not so old as that of vegetable tannages, having been introduced (or more probably re-introduced) by the Moors into Spain, yet the simplicity of its chemistry entitles it to an introductory position. As alum exists in some warm countries as a natural product of the weathering of aluminous shales, it is very possible that its tanning properties may have been accidentally discovered by its substitution for common salt.

Ordinary alum is a crystallised double sulphate of alumina with potash or, more recently, with ammonia or soda; but as the alkaline sulphates take no part in the tannage, simple aluminium sulphate which can now be cheaply made of sufficient purity is often substituted, and from the scientific point of view need alone be considered. The metal aluminium is tri-valent, that is it has three "links" or combining powers, so that the chloride is $Al^{iii} Cl_3$, the oxide alumina, $Al_2^{vi} O_3$; and the sulphate $Al_2^{vi} (SO_4)_3$. It will be found that other metals, and especially chromium and iron, which form salts and oxides of the same type, have also tanning powers.

Sulphate of alumina or alum is always used for leather making (or "tawing") in conjunction with common salt, to the extent of from half to an equal weight, alum alone producing only a stiff and imperfect leather. Various attempts have been made to account for this peculiarity, one being that salt converted the sulphate into the chloride, though unfortunately the chloride *alone* produces no better results. *Basic* salts of alumina, however, will tan without addition of salt, and the true explanation is to be found in the preceding chapter.

A "basic salt" is one in which the base is in excess of the normal proportion to the acid, being partially combined with $-$ OH. Consequently the basic chlorides may be represented by the following scheme :

$$
\begin{array}{lll}
\text{Aluminium chloride} & \dots & \text{AlCl}_3 \\
\text{1st basic salt} \quad \dots & \dots & \text{AlCl}_2\text{OH} \\
\text{2nd} \quad ,, \quad ,, \quad \dots & \dots & \text{AlCl(OH)}_2 \\
\text{Hydrated alumina} & \dots & \text{Al(OH)}_3.
\end{array}
$$

It is doubtful, however, whether in general the basic salts have so simple a constitution, for though some are quite definite crystalline bodies, most are only known in an amorphous or colloidal state, and have all compositions between that of chloride and hydrate, so that for our purpose it is simpler to regard them as colloid solutions of varying proportions of hydrate in the normal salt. Beyond certain proportions of hydrate, varying with temperature and concentration, they readily separate an insoluble basic salt, leaving a normal or less basic salt in solution.

There are various ways in which basic salts can be formed. The most important in theory, if least so in practice, is the direct action of water on the normal salt. It is well known that alum and alumina sulphate solutions are distinctly acid, tasting sour, and reddening litmus. The cause is that a portion of the salt is hydrolysed or combined with water, forming on the one hand actual free sulphuric acid, and on the other a basic salt. This reaction is also common to ferric and chromic salts.

If a skin is placed in an alum solution, it absorbs the free sulphuric acid with avidity and swells. The acid being removed, a further portion of the normal salt hydrolyses, since an equilibrium must subsist between the free acid and the salt, and this hydrolysis continues till the skin has absorbed all the acid of which it is capable at the small concentration corresponding to the hydrolysis equilibrium, and then stops. At the same time the basic salt formed by the hydrolysis is also absorbed by the skin. The acid certainly attaches itself to one or more of the amino-groups (see p. 20) of the complicated skin-molecule : how the basic group is attached is at present unknown; possibly to the COOH or acid-forming group

of the amino-acids. In any case, the two processes
are in a sense independent, and if the salt is originally
basic, a larger proportion can be absorbed than if it
were originally normal, since there can be little
doubt that the introduction of an acid into the skin-
molecule adversely affects its power of absorbing the
basic salt on which true mineral tannage depends.
If common salt is added to the mixture, two effects
are produced, both favouring tannage. As in pickling,
the salt increases the acid absorbed, and at the same
time prevents swelling, and dehydrates and isolates
the fibrils, so producing leather; while the increased
absorption of acid permits the hydrolysis to go further, and thus provides a larger proportion of basic
salt which is also absorbed. Tawing is thus a combination of pickling with true mineral tannage. If
the salt is removed by washing, the acid swells the
pelt, as in a pickled skin, a portion of the basic
alumina salt returns to the normal condition and
washes out; the tannage goes back, and the leather
dries hard[1].

There are several ways by which aluming can be
accomplished. The usual one is to drum or soak the
skin in a strong solution of alum and salt; but sheepskins for rugs, and sometimes fur skins are "cured"
by stretching on a board or frame and sponging with

[1] Other methods of producing basic salts will be explained in
Chapter XIV treating of chrome tanning.

a suitable solution. This method may be recommended for amateur use. The skin is nailed hair down on a board, freed as far as possible from fat and flesh, and sponged with a warm solution of say 1 lb. of alum and $\frac{1}{2}$ lb. of salt in a gallon of water; the liquid is allowed to dry in, and the sponging repeated two or three times till the skin is thoroughly penetrated, when it is allowed to dry, and preferably kept in this condition for a month or more to "age," during which further hydrolysis goes on and the tannage becomes more fixed. The skin is now tawed, but stiff and harsh. It must be slightly moistened by placing in damp sawdust, or in a cellar, and well stretched by drawing it over a blunt edge in all directions, dried more completely, and again stretched, when it should be a very soft white leather. The softening process is called "staking," from the tool used, which consists of a blunt blade with the corners well rounded, which is fixed, edge upwards, on a post. On the large scale the work is now generally done by staking machines.

Such pure alum tannages are little resistant to water, and are also somewhat thin and empty, since only a very small bulk of the alumina salts is absorbed, but they are largely used for belt laces, blacksmith's aprons, whip lashes, and the like. For gloves, means must be adopted to render the leather softer and fuller; and for shoe purposes, more resistant

to water than they would be with alum and salt
only.

The skins suitable for kid gloves are mostly small
lambskins, though real kid is also used. They are
unwoolled by a mixture of lime and realgar, painted
on the flesh only if the wool is valuable; and after
removal of wool and hair they receive a further
liming to plump and soften the pelts. They are
then usually puered and drenched, and are ready
for tawing. This is now done in rotating drums, but
formerly by treading with the feet in large shallow
tubs. Instead of mere alum and salt solution, a paste
is used, containing flour and egg-yolks, and generally
some olive oil, with such a quantity of water that on
taking the goods from the drum the paste remains
adhering to the skins. They are dried hanging on
deal rods, preferably without artificial heat, and after
"ageing" for about a month are damped somewhat,
softened by treading on a "hurdle" or ridged floor,
or by drumming in a dry drum, and are staked.
Before dyeing, they are washed for about 10 minutes
in tepid water to remove flour and surplus alum and
salt, "re-egged" with yolk and a little salt to restore
what has been washed out, and dyed by brushing on
a table[1], first with mixtures of dye-wood extracts
and berries, which are fixed and darkened by a

---

[1] If dyed in a tray, as is sometimes done, the re-egging follows the
dyeing.

second brushing with mordants, such as iron and copper sulphates, potassium bichromate, or (for bright colours) tin salts. After dyeing and drying out, they are again softened by staking or "perching," a process in which the skin is fixed to a bar, and stretched either with a tool somewhat resembling a spade, or with the "moon-knife," an annular and concave disc of steel with a handle across the centre. If skins are too thick, they are reduced on the "perch" with a "moon-knife" with a "turned edge," and the flesh is smoothed and whitened by "fluffing" on a rotating wheel covered with emery. "*Suède*" leather is also fluffed on the "grain" or hair side with a fine wheel. For *Glacé* the skins are rubbed with a "lustre" containing wax and oil, and carefully ironed, and again polished with a second lustre, a little French chalk and a flannel.

The manufacture of the now almost obsolete calf-kid was exactly on the same principle, but adapted to the heavier calf skins. Ordinary limes were often used, sometimes "sharpened" with arsenic, the skins were drenched but not puered; the tawing paste contained more oil, the drumming was much longer, and the skins were usually straightened out and allowed to lie in the paste for 24 hours before drying, while heavy skins were split to save shaving and to utilise the fleshes. Twenty years ago the manufacture was a very large one: one Leeds firm alone used

as much as 50 tons of preserved egg-yolk yearly: but
it is now almost entirely superseded by the grained
and glazed chrome tannages, which are cheaper to
produce and more resistant to water. At present
the fashion for glazed and grained leathers is somewhat on the wane, and it would not surprise the writer
if a chrome leather with the dull finish of calf-kid
again came into vogue.

It has been stated that the object of the flour,
egg-yolk, and oil was to fill and soften the leather.
The starch of the flour is scarcely taken up by the
skin, but the gluten together with the albumen of
the yolk is absorbed under the mechanical action of
drumming or treading, and fixed and preserved by
the salt and alum. The most important part of the
yolk, however, is the egg-oil. The process thus re-
sembles the primitive methods with fat and brains
and kneading, or the more modern " Crown " or
" Helvetia " lace-leathers produced by drumming with
fat and flour, which are mentioned in Chap. XXI.
Other mechanical leathers are produced by all sorts
of modifications and combinations between the alum
and the " Helvetia " processes.

CHAPTER XIV

THE BASIC CHROME PROCESS

IT is quite possible with chrome- or iron-alum (salts of the same type as ordinary alum, but containing chrome or iron in place of aluminium), in conjunction with salt, to produce more or less satisfactory leathers, but these have never been of any commercial importance. The late Prof. Knapp, as early as 1858, published a perfectly practical chrome tanning process, but unfortunately viewed it only as a scientific curiosity, and spent his labour on iron tannages which have never had any commercial success; and it was not until 1893, after Schultz had shown by another method that chrome tanning was of practical value, that Martin Dennis patented it in the United States, almost in the very words of Knapp. Since that time, with slight modifications in detail, it has been in extensive use.

Knapp says[1]: "The acid reaction of iron and chrome salts produces a leather even from thin hides, which is quite too stiff, and liable to break on the grain. If, however, before tanning one adds gradually to the hydrochloric acid solution of the

---

[1] *Loc. cit.* p. 68.

oxide as much sodium carbonate or hydrate as it will
bear without forming a permanent precipitate, one
has the double advantage that the compound of the
oxide is precipitated more easily and abundantly on
the fibre, that the acid reaction on the hide (if not
upon litmus paper) is prevented, and lastly that a
quantity of salt is formed equivalent to the soda
added."

The effect of this procedure is to form a basic
chrome salt. If hydrated chrome oxide be dissolved
to saturation in acid, the solution is already basic,
since a portion of the oxide is dissolved by the normal chromic salt at first formed and the effect of the
added soda is to render it more so, as in the following
equation :

$$2\,CrCl_3 + Na_2CO_3 + 2\,H_2O = 2\,NaCl + CO_2 + 2\,CrCl_2\,(OH).$$

Of course, by varying the addition of soda, any required basicity is obtained. The author showed[1] in
1898 that results in tanning were obtained by the
use of "neutralised" chrome alum, as good as by
the more expensive chromic chloride, and he also
published a method[2] for making a basic liquor from
bichromate, which is still very largely used. Chro-
mates contain the metal in a higher state of oxidation
than the required chromic salts, so that if sufficient

[1] " More about Chrome Tannage," *Lea. Trades Rev.* 1898, p. 400.
[2] " A Cheap Chrome Tannage," *Ibid.* 1897, p. 390.

of a stronger acid is added to liberate the chromic acid and to form a basic salt with the chromium, and half its oxygen is removed, a solution is obtained which will be the more basic the less the excess of acid. The warm solution gives up its excess of oxygen readily to many organic substances, glucose or sugar being generally used. The following equation represents, for the sake of simplicity, the reaction for basic chloride; but in practice sulphuric acid is often substituted:

$$Cr_2O_7K_2 + 6\ HCl = 2\ H_2O + 2\ KCl + 2\ CrCl_2(OH) + 3O.$$

The sugar is largely oxidised to carbonic acid and water, but aldehydes are also produced which remain in the liquid and add to its tanning effect (see Chap. XX), so that the leather is somewhat softer and fuller than that made with basic chrome alum.

The tanning process is a very simple one. The goods prepared for tanning are entered in a weak liquor made by diluting one of the stock liquors above described, which is gradually strengthened by further additions until the goods are tanned. If carried out in a drum or paddle, skins can be tanned in a few hours, but for heavy hides suspension in pits is also a very practical method, though a much slower one. The goods may be moved forward from liquor to liquor, exactly as in vegetable tannage (see Chap. XVII).

When the leather is blue throughout (as may be seen by cutting) it is washed to remove the excess of chrome liquor, and at once "neutralised" in a weakly alkaline liquor. The object of this is to fix and render more basic the chrome salt absorbed by the hide, since it is absolutely essential in the succeeding "fat-liquoring" that no soluble chrome salt should come to the surface. Washing alone would ultimately accomplish this by hydrolysing the salt and removing the more acid portion; it would not only be slow, but wasteful in removing chrome from the skin. "Neutralisation" is a somewhat delicate operation, for while it must necessarily be thorough, it must not be carried so far as to convert the fixed basic salt actually into chrome hydrate, which has no tanning power; and leather treated with excess of alkali dries hard, and is apparently undertanned. Obviously these circumstances exclude the use of "strong" alkalies like caustic soda or even "soda crystals" which even in very dilute solution would carry the neutralisation of the surface too far before the interior was thoroughly penetrated. The measure of the "strength" of alkalies, like that of acids, is their degree of ionisation, but while the property of acidity depends on the concentration of the H-ion that of alkalinity is due to the HO-ion (or hydroxylion)[1].

[1] When an acid is neutralised by an alkali in dilute solution, what really happens is not, as was formerly supposed, the combination

We can, therefore, apply to weak alkalies the
same reasoning which we have previously used with
regard to weak acids for deliming (Chap. VIII). If
we can keep the concentration of the HO-ion low
enough, we can neutralise without fear of going too
far; and, short of this, the weaker the alkali the
more safely will the operation be conducted. For
this reason borax is generally used, about 3 °/₀ on
the pelt-weight in ½ °/₀ solution. Safer still is the
use of whitening (calcium carbonate) which, from its
small solubility cannot carry the process too far, but
in many sorts of leather it is inconvenient.

Probably the best method is that recently published by Stiasny[1], who employs a mixture of
ammonium sulphate and soda crystals. This is
cheaper than but equivalent to employing a mixture of ammonia and ammonium sulphate, as the
soda reacts immediately with the ammonium salt,
setting free ammonia, and the excess of the neutral
salt so reduces the already feeble ionisation of the
ammonia that by varying the concentration and

of the acid-ion and the base-ion ; for as we have seen, the salt
remains ionised and only really combines when the solution is concentrated. The true combination which actually does occur is that
of the H and the OH to form water, which scarcely ionises at all, and
it is for this reason that the actual heat of combination is *the same*
for all acids and bases whatever their character.

[1] Collegium, 1912, p. 293.

proportions the neutralisation may be exactly regulated to the required amount.

The neutralisation is followed by "fat-liquoring," which consists in drumming the skins, which have been previously freed from excess of water by "putting out" on a table with a blunt tool, in a weak emulsion of oil and soap, which is readily absorbed. The oil serves to lubricate and coat the fibres, and the soap reacts with the basic chrome salts, converting them partially into insoluble chrome soaps, and so increasing the softness, fullness and water-resisting power of the leather. If the leather has not been sufficiently neutralised, these chrome soaps are in part precipitated upon, instead of inside the leather, and render its proper dyeing and finishing impossible.

CHAPTER XV

THE TWO-BATH CHROME PROCESS

This process, invented by August Schultz in 1884, was the first really successful one, though a combined tannage with chrome and alum had been earlier patented by Heinzerling with more or less practical result. Schultz was not connected with the leather trade, but was a dyers' chemist; and it is said that his attention was drawn to the subject by a friend, who wanted a leather to cover corset-steels without the tendency to rust caused by alum and salt. About the same date he had been interested in a process for mordanting wool by reducing bichromate on the fibre, and conceived the happy idea of applying a similar method to leather.

Schultz first treated the skins with a weak bath of potassium bichromate acidified with hydrochloric acid (or practically with dilute chromic acid), and this was absorbed by the skins, colouring them bright yellow, but producing no tanning effect. They were then transferred to an acidified bath of sodium thiosulphate ("hypo"). The free thiosulphuric acid is a powerful absorbent of oxygen, and immediately reduces the chromic acid in the skins to a green basic chrome salt, which at once converts them into leather, while the thiosulphate undergoes a complicated oxidation to sulphate and tetrathionate, and at the same time deposits a good deal of free sulphur in and on the skin. As thiosulphate is always used in excess of the acid present, it also effects a considerable amount of "neutralisation," as thiosulphuric is a somewhat weak acid. Usually, however, a further neutralisation with borax is practised, and the goods are then fat-liquored and dyed before drying, for it is an inconvenient peculiarity of chrome leathers that once they have been thoroughly dried,

it is impossible to bring them back into such a
condition that they will dye freely and easily. The
chromed skin, like a piece of cloth, is easily penetrated, and even saturated with water, but the actual
fibres do not swell or absorb freely. The process
produces a very soft and pliable leather, and is still
in considerable use, mainly for glazed goat, but also
for box-calf, with no change in principle and very
little in detail from that originally patented by
Schultz. Probably one of the causes of the softness
of the leather is the complete absence during the
process of any acids stronger than chromic or thiosulphuric, but a still more important one is the
deposition of sulphur in the skin and on the fibre,
which acts as a filling and fibre-isolating substance,
and also combines with the oils employed in fat-
liquoring. It is possible to produce an actual leather
by a sulphur-tannage alone, by neutralising a pickled
and consequently acid skin with thiosulphate or a
polysulphide, such as "liver of sulphur," and skins
tanned by the simple basic chrome process may be
made closely to resemble those of the two-bath, by
employing one of these salts to "neutralise" the
somewhat acid leather. Free sulphur is easily detected in leather by wrapping up in a small piece
of paper with a silver coin, which will blacken in a
few hours, especially in a warm place. For the same
reason, two-bath leathers in imitation of chamois

are unsuitable for cleaning silver, but basic leathers are free from this objection and have good polishing properties.

In order to get still greater softness than is attained by the chrome processes alone, skins are frequently pickled with salt and acid, or salt and alum before the tannage proper. If alum is used the resultant leather is a combination of chrome and alumina tannage. Sulphate of alumina is also sometimes added to the bichromate which renders it basic, its acid liberating at the same time a portion of chromic acid from the bichromate. It is possible to chrome a pickled skin in a bath of bichromate alone, as it contains in itself sufficient acid to decompose the bichromate, which at the same time "depickles" the skin, so that no salt or alkali is needed.

Neutral skins fix scarcely any chrome in a bichromate bath without acid, and as the bichromate is usually employed in excess, the quantity of acid used determines the extent of the chroming. Sulphuric acid is often substituted for hydrochloric in both chroming and reducing baths, the only difference being that sodium sulphate instead of chloride is formed, while if common salt is desired it is easily added. Of course sodium bichromate may be substituted for that of potash, and practically weight for weight.

The finishing of chrome leathers is a somewhat

complicated process. For box-calf the skins are "put out" or stretched, either on the table with a "slicker" (sleaker) (Fig. 13), or with a machine; partially dried ("sammed"), and shaved or split before dyeing. Black dyeing is done either on the table by brushing, or in a vat, by alternate treatment with logwood and iron solutions, often assisted by some coal-tar dye; and colours may be produced by hot dyewood solutions followed by appropriate mordants, but only the "acid" coal-tar colours will dye on chrome leather direct, the "basic" colours requiring a previous mordanting with tannin. This indicates a profound alteration in the chemical affinities of the hide fibre by chrome tannage, since the raw skin absorbs basic colours with the greatest avidity.

After dyeing, the skin is dried wholly or partially, damped back if necessary and softened by "staking" on a machine. It is then "grained," if for "Box calf," by folding the skin grain in, and rolling the fold under a cork-covered board on a table. This produces minute longitudinal creases on the grain surface, and by crossing these, the well-known checked pattern of this sort of leather is produced. The operation is not easy to describe, but will be again alluded to in the chapter (XIX) on Moroccos, *q.v.* In glazed kid this treatment is omitted, as a perfectly smooth surface is desired.

The skin is now completely dried, and the surface

is very slightly moistened with a "seasoning" con-

Fig. 9.   Glazing machine.

sisting mainly of a weak solution of egg- or blood-albumen with a little milk, and often some dye ; and as soon as the moisture has sunk in, it is glazed by friction under a smooth cylinder of glass or agate held in the arm of a glazing machine (Fig. 9) which is pressed heavily on the leather as it moves away from the workman, but lifts on the return stroke. As this somewhat hardens the leather, it is again softened by staking, and, if necessary, glazing and staking are repeated till the necessary gloss is attained.

Glazed chrome kid is chiefly an American product, though its principal raw material is dried Indian goat. The manufacture originated in the States and is carried on on an enormous scale, more than one of the Philadelphia factories having a capacity of three or four thousand dozen a day; and though some few firms in England can produce leather of quite equal quality, competition is almost hopeless so long as these are satisfied with a few hundred dozen a week.

CHAPTER XVI

THE VEGETABLE TANNING MATERIALS

THE tannins may be described as a class of substances found in many plants, which have the common properties of precipitating gelatine from solution and of converting skin into leather. They are all

7—2

colloid, that is uncrystallisable, and for this reason few of them have yet been obtained in a pure form. They are feebly acid, and are sometimes called "tannic acids," but it is uncertain whether they are strictly acids, since many phenols, of which "carbolic acid" is a type, have also slightly acid properties. All natural tannins[1] are benzene derivatives, either from the dihydric phenol catechol, or the trihydric phenol pyrogallol, and another trihydric phenol phloroglucol is also often present. The positions of the OH groups are shown by the following diagram, the carbon atoms at the other angles being combined with H.

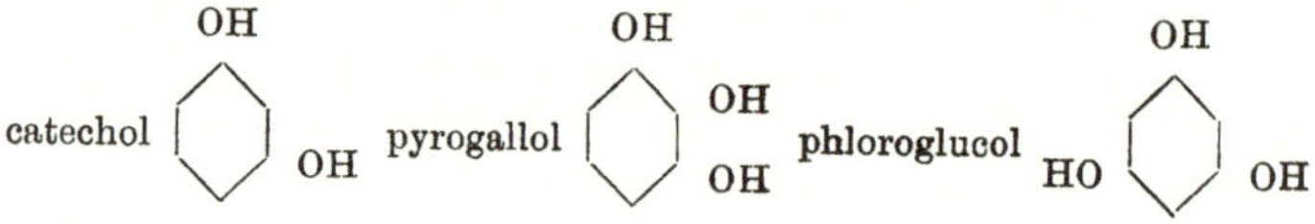

They are consequently usually divided into "catechol" tannins giving green-blacks with iron salts, and "pyrogallol" tannins giving blue-blacks, and often used as inks; but it is certain that the distinction lies deeper, and is rather one of structure than of the particular phenol. The two classes, however, whatever the cause, possess a marked difference in tanning properties, the iron-blueing tannins

[1] Fuller information as to structure is given in the *Lea. Ind. Laboratory Book*, Chap. XIX.

causing a white deposit of crystallised ellagic acid
in the leather while the iron-greening (with a few
giving violet-blacks) deposit dark brown substances
called "reds" or "phlobaphenes." Whether these
are in all cases products of the tannins themselves,
or rather of other bodies associated with them, is
still doubtful.

The tannin-yielding materials are so numerous
that but very few, even of those in commercial use,
can be mentioned here. The oldest and formerly
the most important in this country is the bark of
the oak, usually stripped in spring when the sap has
begun to rise, because at this time the bark is more
readily separated from the trunk. It is a somewhat
weak material, only yielding $10—12\,°/_\circ$ of substances
absorbable by hide, but it has the peculiarity that
good leather of almost all descriptions, both light
and heavy, can be made by its use. Very little
leather is now tanned exclusively with oak-bark,
though other oak products are largely used. Oak
wood also contains tannin, though in less quantity
than the bark; but by chipping and hot extraction,
and subsequent decolorisation and concentration of
the infusion by evaporation, an extract of $25—30\,°/_\circ$
is obtained, and is now made from the waste wood
and sawdust in very large quantity in Slavonia and
Northern Italy, where oak is still abundant. An
extract hardly to be distinguished from that of oak

wood is made from the wood of the edible chestnut
Castanea vesca.   The oak principally used in Europe
is *Quercus robur* (with its sub-species *Q. sessiliflora*
and *Q. pedunculata*) but most oaks contain tannin.
Very important is the large bearded cup of the
acorn of evergreen oaks from Greece and the Levant
known as valonia, which is extremely rich (30—40 $^\circ/_\circ$)
in tannin, and largely imported.   Oak-galls (from
Q. infectoria) are a source of the druggists' "tannic
acid," but unimportant as a tanning agent, though
the often repeated statement that the so-called
"pathological tannins" from insect galls will not
make leather is absolutely without foundation.   It
is somewhat curious that these oak tannins are by
no means chemically identical.   The bark-tannins
distinctly belong to the class we have called catechol
tannins, though that of the common oak contains
some mixture which gives blue-black with iron and
produces "bloom."   The wood tannins belong to the
pyrogallol class, and the gall tannin is the typical
"gallotannic acid" which is a pure pyrogallol derivative.

Another important material of the pyrogallol class
is myrobalans (30—40$^\circ/_\circ$), the dried fruit of an Indian
tree, and divi-divi, the pod of *Cæsalpinia coriaria*, a
tree allied to logwood (40—50$^\circ/_\circ$) and sumach, the
leaves of *Rhus coriaria* (25—30$^\circ/_\circ$), while the most
important sumach adulterant, *Pistacio lentiscus*, is a
catechol tannin.

To the catechol class belong also gambier, an extract from the leaves of *Uncaria gambia* (30—40°/$_{\circ}$); the bark of the Australian mimosa or "wattles" (25—40°/$_{\circ}$); quebracho extract from the very hard wood of a South American tree (dry 60—70°/$_{\circ}$) and many others. It will be noted that all these materials are very much richer in tannin than oak-bark, and it is to the stronger liquors obtained from them, rather than to any "chemical tanning," that the shortened time of modern sole-leather tannage is due. Whether the shortened process gives so durable a leather as the older method may be questioned, but the leather is honestly and thoroughly tanned with mixtures of natural tannins very closely approximating to that of oak-bark; while in the old process much time was positively wasted by ignorant mismanagement.

The way in which tannins tan and the chemical nature of the leather formed still need elucidation. The view most probable at present is that leather is rather a colloidal than a strictly "chemical" compound. The tannins all yield colloidal solutions, and gelatine and hide-fibre are typical colloids. It has been shown in the author's laboratory and elsewhere that, in presence of the trace of acid essential to tanning, the particles of tannin have opposite electrical charges to those of gelatine, and it is well known that two colloid solutions in which this is the case, when mixed, are mutually precipitated as a colloidal compound.

Although not strictly vegetable materials, the synthetic tannins recently discovered by Dr Stiasny must be mentioned here, as they will probably have considerable commercial importance. They are coal-tar products, produced by the condensation of sulphonated phenols with formaldehyde, and produce almost perfectly white and very soft leathers. Though not identical with any natural tannins, they possess most of the characteristic properties of the class, precipitating gelatine and basic dyes and giving blue-black inks with iron.

CHAPTER XVII

THE VEGETABLE TANNING PROCESS

THE tannage of sole-leather, though full of problems and difficulties in commercial practice, may in theory be regarded as one of the simplest; and a clear understanding of its methods will render easy the comprehension of the rest.

The preparation and character of hides for sole-leather have been described in Chaps. VI and VII. To recapitulate, they are somewhat rapidly limed in

fresh limes generally somewhat "sharpened" with
sulphide of sodium; the whole process being so
conducted as to produce good swelling and easy
unhairing, with the least possible solution of valuable
hide-substance which should go to make a heavy and
solid leather. After unhairing and fleshing, the hides
are "rounded" or trimmed, as only the thicker and
central part called the "butt" is suitable for soles.
After washing in water the butts were (and in some
tanneries still are) considered ready for tanning, but
it is becoming more and more usual first to remove
the surface-lime by a bath of boracic or some other
weak acid, not only to secure a brighter and more
uniform colour, but to economise the natural acids
of the tan-liquors, which in the modern rapid process
are much less freely formed than in the old method,
in which there was abundant time and material for
bacterial and other fermentations.

The butts first go into "suspenders," a set of 8 or
10 deep pits in which they are hung by string to
sticks laid across the top of the pit. Sometimes
these sticks are supported on a frame to which a
gentle reciprocating motion is given by suitable
machinery, which causes a constant flow of liquor
between the butts and prevents their remaining in
contact, which would cause stains; but otherwise
they must be moved and shaken by hand, especially
in the earlier stage of the process.   The pits are

generally arranged so that liquor can flow from the top of one pit to the bottom of the next, so that the fresh liquor is all pumped into the strongest pit, and flows away exhausted from the weakest into which the "green" butts from the limes are brought and moved forward daily in the opposite sense to the liquors, which are appropriately the oldest and most exhausted in the yard; though if fresh liquors must be used, gambier or myrobalans are among the most appropriate materials. The object of using the oldest liquors is not merely one of economy, but because such liquors are what the tanner calls "mellow"; that is, their action on the hide is gentle, and only mildly astringent. This arises from several causes. Even if only one material is used, the tannins contained in it are a natural mixture; and the ordinary liquors of "mixed tannage" contain a still larger variety, varying much in their affinity for the hide-fibre. It is obvious that when a liquor is brought in contact with partially tanned hide, those tannins which are most astringent, and have the greatest affinity for the fibre must be removed first, so that what remains at last is only the mildest and least active part. Most tanning materials also contain a considerable portion of what the leather-chemist styles "non-tannins." These are partly sugars and other carbohydrates, which during the process are gradually fermented by bacteria and yeasts to organic acids; and partly derivatives of the tannins themselves which, though they do not actually tan, are yet to some extent absorbed by raw hide, and promote its conversion into leather.

We may add to this the gradual accumulation in the used liquors of organic salts of lime and potash which weaken the acidity of their corresponding acids, and so keep the tanning action in check, since a certain degree of acidity is necessary for the process, and sufficient addition of alkaline salts may even bring it to a standstill.

The first process which takes place in the suspender is the neutralisation by its acids of any lime which still remains in the hide, and the consequent reduction of alkaline swelling. The hides, which, if not previously delimed, were at the outset plump and elastic, become very soft, and easily impressed by the finger; but in a properly conducted process should not lose much actual thickness, since the liquid which escapes from the fibres remains between them, and, as the lime is gradually replaced by the weak acids of the liquors, the fibres again swell and the hide again increases in firmness. That this should take place is essential to the production of a firm and solid sole-leather, for pelt tanned in the flaccid and "fallen" condition remains soft and porous, as well as thin. Fibres which have once been swollen by lime are much more sensitive to acid swelling than where this has not been the case, and those unhaired by "sweating" or any other process in which swelling has not occurred, require much stronger acids to produce adequate swelling and differentiation of the fibre bundles. For this reason American "sweated" sole-leather is usually swollen by dilute sulphuric acid in an early stage of the process, when the surface only has been tanned and thus rendered insensitive to the action of acids. If applied without this precaution, mineral acids produce a dark and brittle grain, and though this is prevented by the slight tannage, a dark layer may usually be detected beneath the thin tanned surface. For this reason, if "acid" leathers are "buffed" or glass-papered in shoe-manufacture, the surface is usually blacked and burnished or covered with a coloured "fake" or paint to conceal the actual leather and imitate a better material.

At the same time that the alkaline is being replaced by the acid swelling, the surface of the leather is taking colour, most rapidly indeed if the hides are somewhat limey; and actual tannage gradually penetrates, so that by the end of the suspender period, the hide, if not tanned, should be nearly or quite "coloured through."

At the end of the suspenders the butts, still soft, are advantageously laid flat to straighten out any lumps or creases which may have formed, before

they go to the "handlers." Though the time of
suspension is short and the liquors weak, the latter
must be abundantly supplied, for the green hides
have great avidity both for tannin and acid, and if
allowed to remain in exhausted liquors, putrefaction
sets in, which brings down the hides as in a bate,
and renders it impossible later to produce a firm and
plump leather. Too great astringency, on the other
hand, is equally dangerous, since it also checks swelling, and hardens the surface, producing perhaps a
"semi-permeable membrane," which it is very difficult for the tannin to penetrate later. Such liquors
also cause "drawn grain," for if the surface of the
hide is fixed by too early tanning, and the interior
afterwards swells in thickness and contracts in area,
the grain becomes puckered in wrinkles which are
the larger and coarser, if through ill-managed liming
the fibrous texture immediately below the surface is
too much loosened.

The "handlers" are a series, conveniently say of
12 pits, in which the butts are laid flat instead of
suspended, since in this way a much larger number
can be got into each pit; and the hides and not the
liquors are changed from pit to pit. The "pack"
from the suspenders is, of course, brought first into
the weakest and oldest pit; the "forward" or most
tanned pack is removed to the "layers," or in some
cases to a second and stronger "shift" of handlers;

and the intermediate packs are each moved forward
one pit, all getting thus a change into a stronger
liquor.   As, however, the strength of the liquors
must be maintained by fresh liquors, if possible from
the "layers," the new "forward" pack is again raised
(conveniently next day), the whole series is again
shifted forward, and the liquor vacated by the last
pack is run to the suspenders, and replaced by a new
liquor to receive the forward pack.   If this and the
bringing in a new pack takes place on alternate days,
the passage of a pack through the entire "shift" or
"round" will occupy four weeks, including Sundays.
It is most usual to add to the liquor of the forward
pack a few pailfuls of finely ground material, bark,
myrobalans, or valonia.   The object of this is less to
increase the strength of the liquor than to prevent
the actual contact of the butts.   Before bringing the
later packs into a pit, this "dust' is well plunged up
to distribute it as evenly as possible between the
butts.   It is not uncommon to have two or even
three head packs or "dusters" which are moved
alternately so as to give them a longer time in the
new liquors, but this introduces complications which
cannot be discussed here.

   In the handlers and layers the goods were formerly handled with sharp steel hooks on long poles,
but as these are apt to produce serious scratches
they have been to a large extent abandoned for the

method of handling with strings, described on p. 27, which is also much quicker, and avoids the need of partially skilled labour.

If the liquors are well maintained, and the goods are light, they will probably now be ready for the "layers," that is to say, thoroughly tanned through, but still wanting in solidity and weight. They are no longer capable of rapidly absorbing tannin, but can still fix considerable quantities of bloom and reds (p. 101) on and between the fibres. These difficultly soluble matters are supplied by strong and fresh liquors which are supersaturated, and deposit them not merely on and in the hides, but in the liquors and on the sides of the pits. It is hence desirable to bring the actual tanning materials as closely as possible in contact with the leather, and to maintain the strength of the liquors during the comparatively long periods for which the butts must remain undisturbed. To accomplish this, much larger quantities of strong solid materials, such as valonia, mimosa bark, and (in moderation) myrobalans are employed than in the handlers, and these serve also the purpose of separating the butts and maintaining a larger volume of liquor between them. Instead of throwing this material in three or four portions into the liquor, and trusting to its distributing itself between the goods, each butt as it is drawn on to the surface of the liquor, receives a regular sprinkling;

a second is drawn on and similarly sprinkled, and so
on, and the whole are allowed to sink as evenly as
possible into the liquor. This method of making a
layer is now universal in England, but an older
method is still largely in use on the Continent, in
which the goods (generally whole hides) are spread
in the vats with thick layers of dry material between
them, and the liquor is only run on when the pit is
filled, which is then often left undisturbed for three
months or more. The method dates from a period
when the solid materials rather than the "weak"
liquors were relied on to supply the tannin, and has
the disadvantage that the hides are often deeply
pitted by fragments of tanning material pressed
into them, but it must be admitted that the leather
produced in this way, though sometimes unsightly, is
very solid and durable.

The liquors used for the last layers are the
strongest which can be made and, since the almost
universal use of oakwood and chestnut extracts,
frequently reach 100° or even 120° of the barkometer
(*S. G.* 1·100—1·120) while in pure oak-bark tannage it is difficult to exceed 30°, and then only by
strengthening repeatedly used liquors. The goods,
when it is desired to change the liquors, are pulled
out, the liquor pumped off and the partially exhausted
solid material removed for further extraction, and
the goods are returned to a new liquor in the same

pit. Layers of a month to six weeks were formerly the rule, but it is better practice to begin with one even as short as a week or ten days, gradually extending the time as the goods become more fully saturated, and exhaust the liquors more slowly. It is only by constantly maintaining and increasing the strength that the goods will continue to "feed," and it is waste of time to allow them to lie in a liquor which has fallen below its original strength. Of course, if layers are to be short, less solid material is given. Light goods are fully tanned in two or three layers ; heavy ones may require four or five, but the careful tanner will investigate how far the gain of weight and solidity repays for the cost of additional time and material, which, as the goods become fully tanned, produce but little effect.

After the last layer, goods were formerly merely washed in a weaker liquor before being sent to the drying shed, but some sort of bleaching is now almost universal where extracts are largely used. A strong warm myrobalans liquor produces a good effect, but recently "vatting" in warm strong solutions of "bleaching extracts" after scouring has become customary. These are mostly quebracho extracts containing much bisulphite of soda, which not only bleaches by its sulphurous acid, but has the property of dissolving and removing the "reds" deposited in the leather. The effect is to render the leather brighter in colour without removing much weight, but at the same time to make it more porous and much more permeable to water. It is hard, however, to see how the practice is to be avoided so long as the public and the shoe-manufacturers continue their absurd demand for soles of light and bright colour !

The finishing of sole-leather is simple in theory, but, especially with heavy extract tannages, not at all

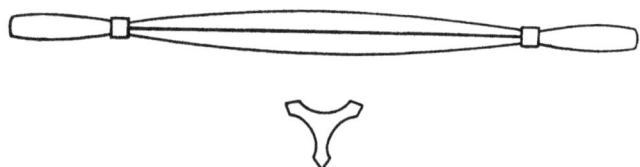

Fig. 10.   Striking pin and section.

easy in practice. Before drying, the goods are very lightly oiled on the grain side, usually with crude cod liver oil, with the object both of lessening oxidation by the air, which darkens colour, and of checking evaporation from this side, which would tend to bring the dark liquors, still present in the interior, to the surface by capillarity. The first drying must be slow and even, and at a low temperature, and if possible is accomplished without artificial heat. When the goods are half dry (or "sammed") they are laid in a pile, damped in dry places, and often allowed to heat a little by the incipient growth of mould, which, though somewhat dangerous, facilitates the next process, that of "striking." This, if done by hand, is accomplished on a horizontal wooden beam of rounded section on the upper surface, and about 7 ft. long. The tool used is called a "striking pin," and is a two-handled blade of triangular section with three edges, with which the workman stretches and smooths the grain side of the leather, leaning over the low beam, and putting his full weight on the tool. If it is desired to strike the bloom "out," so as to show the original colour of the leather as fixed in the suspenders, the surface is kept thoroughly wet, often with slightly soapy water, the tool used is pretty sharp, and is aided by a stiff brush. If the bloom is to be struck "in," and afterwards concealed by colour, a blunter tool, a dryer leather, and a little oil are employed, and loose bloom is afterwards washed off. The process, however, is now generally performed by a machine carrying four blades or "slickers" on separate spring-arms which work outwards from the centre of the butt, and lift on the return stroke (Fig. 11). In place of removing the bloom by striking it is now generally removed by scouring tools of smooth sandstone, aided by brushes and plenty of water, before vatting and usually by machine. After the first striking, the goods are

8—2

dried a little further, piled again overnight, and again struck to smooth them and remove tool marks, often

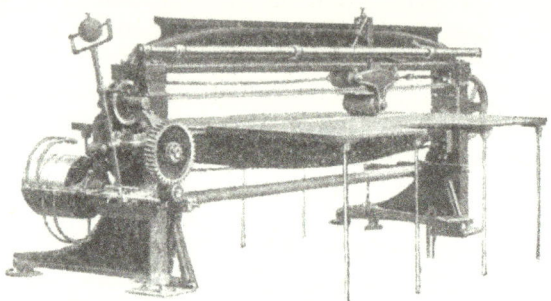

Fig. 11.   Butt rolling machine.

after rubbing the surface with an oily rag.   After the second striking, they are rolled, now almost univer-

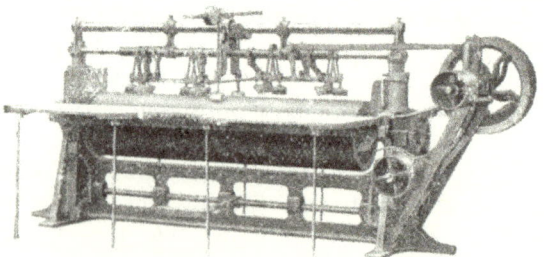

Fig. 12.   Wilson's striking machine.

sally by machines of the type shown in Fig. 12, coloured if desired by sponging with a pigment or

dye solution, often containing some mucilaginous
ingredient and a little linseed oil to heighten the
gloss, again rolled with a heavier pressure, and hung
up in a warmed and well-ventilated room to dry off.
The whole process of finishing takes nearly a fort-
night, and much judgment is needed as to the exact
"temper" or degree of dryness required for each of
the operations. Such is a brief description of the
usual methods, which, however, vary in detail in
different yards.

The common American method is much simpler
and rougher. The goods are dried completely after
light oiling, damped back and piled to equalize, and
rolled twice with pendulum rollers driven at a high
speed, which give a good gloss, but not the even
colour and freedom from wrinkles which is demanded
in English leather.

CHAPTER XVIII

CURRIED LEATHERS

Curried leathers include a large variety of products from heavy strap butts to calf skins, and we
must content ourselves with a few remarks and a
typical example only. The tannage of strap butts
differs little, and often not at all from that described

in the last chapter, but for shoe upper leathers, where a really soft material is required, it is much modified both in the tannage proper, and in the preparation of the hide, which must come into almost non-acid liquors in a thoroughly bated or flaccid condition, since any swelling during the process produces a hard leather.

As a typical example, the vegetable tannage of East India kips may be taken, for though it appears to be a dying industry, it illustrates in a marked way the differences between sole and dressing leathers. East India kips are the hides of small Indian cattle (see p. 5) which arrive here either simply dried, or after a preliminary treatment with a soupy mixture of a salt earth, which adds to weight but also helps to preserve the kips, and render them easier to soften ("plaster kips").

The first treatment is the soaking, formerly done in putrid water rarely changed. This dangerous process is now superseded by the use of fresh water for each pack, with the addition of small quantities of caustic soda or sulphide of sodium; so that "stocking" or pounding in a fulling mill is no longer needed, and is replaced by drumming with a little tepid water. Probably treatment with dilute sulphurous or formic acid is a still better method.

The thoroughly softened goods receive about a month's liming in rather old limes, with the object of still further softening, even at the cost of some loss of hide-substance; and some sulphide of sodium is now usually added. The kips are often unhaired by drumming or stocking, and are fleshed on the beam in the usual way, or sometimes by machine. They are then bated for four or five days in a cold infusion of hen or pigeon dung, best made by extracting the dung in vats, and allowing the clear yellowish liquor to ferment a day or two before use. If the skins are still not sufficiently soft, bating is supplemented by a bran drench. The goods are now quite "fallen" and flaccid; and after "scudding," or working on the beam with a blunt knife to remove remaining hair-roots and fat glands, are ready for tanning.

The "colouring" is done in a weak old handler liquor in a paddle-vat, and occupies at most a few hours, after which the goods are taken at once to a handler "round" and the liquor used in the paddle is run away.

The handlers are worked precisely as in sole-leather, but the liquors are weaker, and very "mellow," being rarely run away (except in the case named) but returned to the leaches and used again and again, and from the accumulation of neutral salts and the small amount of acid they contain have practically no swelling action. The materials of the ordinary Leeds tannage are gambier and valonia,

often with addition of mimosa bark, and myrobalans.
It seems curious that valonia, which in sole-tanning
gives a hard and heavily bloomed leather, should be
used for a purpose in which soft leather and absence
of bloom is particularly desired and obtained. The
explanation of the paradox lies in the method of use.
For sole-tanning, valonia is used for dusting, or if for
liquors, these are made at a low temperature, and
are run at once on the layer-pits. For dressing
leather the valonia is extracted hot, and the liquors
are strengthened repeatedly, a little fresh valonia
being added to the extraction vat, or "leach," for
each liquor. These conditions promote the deposition
of bloom in the leaches and on the tanning material,
and little of the bloom-giving tannins reach the
leather, which is very frequently handled, so that
bloom has not time to deposit. No "dust" is used
in the handlers; and the forward packs, answering
to the "dusters" in sole-leather are called "liquor-
packs." The tannage is almost completed in the
handlers, and the goods, instead of regular layers,
are merely laid away in strong liquors without
"dust" for two or three weeks. At the end of the
tannage the colour is brightened by drumming in
warm strong sumach liquor containing the ground
sumach, which acts to some extent as a scouring
agent, and is washed off later in clear liquor or
water; and the goods are then piled three days to

drain, well oiled, and taken into the shed to
dry out.

In bark yards the process is more like that for
sole-leather, but with weak, and often old liquors;
and the goods receive one or two layers with ground
bark between them. A good deal of bloom is therefore deposited, which has to be scoured out by the
currier.

Currying is to a large extent a mechanical operation, cleansing, reducing in thickness and softening
the leather; the chemical part of the process being
the impregnation with oils and fats. Many of the
peculiarities of the English trade are the result of
bygone legislative interference, which is almost always prejudicial to business. It was formerly illegal
to carry on together the two trades of tanning and
currying; and thus two operations, which were
naturally parts of one process, became separated;
and leather was dried out by the tanner to be wetted
again by the currier, instead of proceeding at once
to curry the wet leather, as is usual in America and
on the Continent, and increasingly so in England.

Whatever be the leather, it is almost invariably
soaked and softened in sumach infusions as hot as
the arm will bear, and is then laid on a table of
marble or mahogany, and thoroughly scoured with
brush and slicker, and plenty of water, till the bloom
(if any) is entirely removed. In many American

cheap tannages actual scouring is dispensed with, the tannage being so conducted, and with such materials (mostly bark of the hemlock fir, *Abies canadensis*) as to entirely avoid bloom. The "slicker" is simply a flat steel blade with a square edge about six inches wide by three or four deep, set in a wooden backing or handle which can be grasped with both hands, but the work is now mostly done by machines similar in principle to the "striking" machine (p. 116), though usually working on flat movable tables.

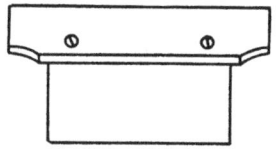

Fig. 13.   Slicker.

After scouring, the goods are partially dried (sammed), frequently by pressing or squeezing between rollers, and are then shaved or split. Hand-shaving is done on an almost vertical beam faced with glass or lignum-vitæ, with the curriers' "head-knife," but its place has been largely taken by shaving and splitting machines. The head-knife, however, is so curious and ingenious a tool as to deserve description.

It consists of a heavy rectangular two-edged blade about 10″ long by 6″ deep, of soft steel, strengthened

and held by a bar down the centre which carries a
handle at each end, the left in the line of the bar and
the right-hand one across it.   The knife is first ground

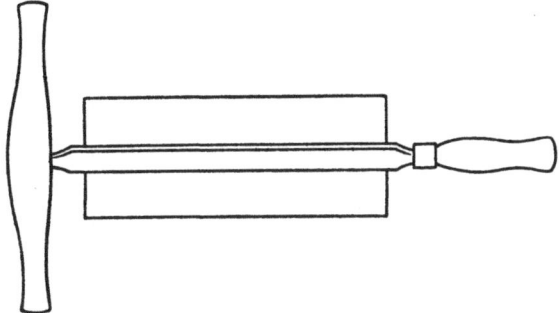

Fig. 14.   Curriers' shaving knife.

to a sharp edge and set on a "Water-of-Ayr" stone
in the ordinary way, and is then placed between the
knees with the straight handle and one end of the cross

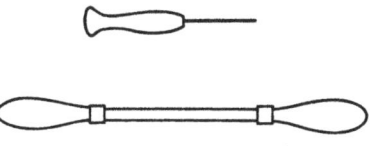

Fig. 15.   Curriers' steels.

one resting on the ground, and the edge is gradually
turned over at right angles to the blade by rubbing
with heavy pressure with a smooth "steel" held with

both hands. Such a "turned" edge will cut thin
shavings off soft leather, or even off a soft "bated"
hide. A skilful workman will keep his edge in order
for a long time with a little steel, almost like a piece
of knitting needle in a handle, which is passed over
the edge, first outside, and then down the angle
formed by turning. The shaving machine uses a
spiral rotating knife, which also to some extent has
a turned edge, caused by an emery wheel rotating
faster than the blade spiral, and thus throwing the
"wire" forward. The splitting machine deserves a
chapter to itself, for it has revolutionised light-
leather manufacture, but must be dismissed in a few
words. That most commonly in use, the "band-
knife" machine, consists of a steel belt about 2″ wide
running rapidly in tension over two pulleys through
a horizontal grooved guide, and kept sharp on one
edge by an emery-wheel. Against this edge the hide
is forced by passing between brass rollers, the upper
of which is straight and perfectly rigid, while the
under one is in sections resting on a lower roller of
india rubber and can thus give to the inequalities of
the leather. The thickness of the split is regulated
by raising or lowering these rollers, and an upper
split can be taken, almost as thin as paper if desired,
and quite uniform; while the inequalities remain on
the lower split, which is often levelled by splitting a
second time.

After shaving or splitting, comes the "stuffing"
or impregnation with grease. The old method, still
in use, is to coat the damp leather on one or both
sides with a thick layer of "dubbing" (originally a
mixture of almost equal parts of cod oil and tallow)
and allow it to dry slowly at a moderate temperature.
As the water evaporates, the liquid part of the
dubbing takes its place by capillarity, becoming
thoroughly distributed over the fibres; and a layer
of the harder fats, called "table-grease," is left on
the surface and is afterwards removed by slicking.
The object of the hard fats is thus mainly to give
coherence to the mixture, and enable it to adhere
to the leather in sufficient quantity; what is really
absorbed by the leather being a saturated solution,
principally of the softer fats of the tallow, in the fish
oil. The oil itself has the further function of per-
forming a sort of secondary tannage, comparable to
the primitive fat-tannages which have already been
mentioned. This property is most marked in the fish
oils, which are partially "drying" oils, containing
large quantities of unsaturated fatty acids capable
of absorbing oxygen and becoming converted into
resinous bodies on the fibre. Excess of such oxidisable acids leads to a troublesome complaint known
as "spueing," in which little warts of these sticky
resinous matters appear on the surface of the
leather.

The modern method, known as " drum-stuffing,"
is universally applied to the cheaper leathers, and is
not only much more rapid but incorporates larger
quantities of the harder fats. While about 10 °/$_\circ$
of total fats is all that can usually be got in by
hand-stuffing, some drum-stuffed goods (especially
the American) contain nearly 50 °/$_\circ$ of greases.

In drum-stuffing the wet leather is introduced
into a " drum " or rotating barrel, heated by hot air
or steam to about 140° Fah., and after rotating for
a short time to warm the leather, liquid grease at
about the same temperature is introduced through
a hollow axle and drummed into the leather. The
whole process is complete in about half an hour, the
goods being thrown on the floor to cool, and set
out on a table with a slicker while still warm and
flexible. In drum-stuffing the fats are often mainly
hard greases, pressed tallow (" oleostearin "), distilled
stearine (crude stearic acid from recovered greases),
and often paraffin wax, with some sod or fish oil, and
this is particularly the case where light-coloured and
dry-feeling leather is desired.

After drying, the flesh is usually " whitened " with
a very peculiar slicker with a turned edge, to remove
surplus grease, and smooth the surface. This process
has practically superseded shaving at this stage with
the turned edged knife. Sometimes a thin film of
the grain is also taken off with the same sort of tool

("buffing"), principally to remove defects, such as barbed wire scratches.

The subsequent finishing processes differ much with the kind of leather. For "waxed" goods (calf, kips, and splits) a grain is raised by "boarding" (p. 132) with a cork board, generally first lengthways and across, and then from corner to corner, so that a "pebbled" grain of no definite form results. This process also softens the leather, and, if hard greases have been used, very much brightens the colour, by breaking up the grease into a powdery state and separating the fibres.

The blacking of waxed goods is done on the flesh side, in England with a mixture of lamp-black and oil, and in America principally with lamp-black and soap, and is followed by a "bottom size" containing much glue and some tallow, and afterwards by a "top size" of somewhat similar character, both of which are smoothed and brightened by "glassing" or rubbing with a thick smooth glass slicker. In the States a "paste" of flour and soap is substituted for the bottom size, and the finishing size is similar, but with the addition of gum-tragacanth.

Leathers in which the grain is blacked, and worn outside, are generally brush-dyed with logwood and ferrous sulphate. The grain is raised by boarding, but this is usually assisted by first "printing" with an engraved roller in an embossing machine, and many of the cheaper leathers are simply printed, often with a grain to imitate some better material. Even the flesh-splits, which have no natural grain, are frequently sized and printed in this way; and more recently an artificial grain has been formed with a celluloid varnish similar to the "new skin" of the druggists, which makes a waterproof and fairly durable leather much used for the cheaper sort of men's boots.

CHAPTER XIX

MOROCCOS AND FANCY LEATHERS

GENUINE morocco is goatskin, or hair-seal. Possibly the latter is not strictly entitled to the name, but as, for many purposes, it is even superior to goat, and is always sold under its own name, it can hardly be called an imitation. Sheepskins are often finished as morocco, and some breeds of coarse-woolled sheep are difficult to distinguish from the genuine article. This is specially the case with the Kazan and other Eastern sheep (see Chap. II).

The tannage of the best moroccos should always be with sumach, but large quantities both of goat and sheep (so-called "Persians") are tanned in India with the bark of *Cassia auriculata*, a bush closely allied to senna. This produces a very pretty leather, well adapted for slippers and other articles of daily wear; but it is quite unsuitable for bookbinding and upholstery which have to withstand the effects of light, and often of gas-fumes for long periods. Under these influences the cassia leather becomes red and tender, and the surface crumbles, and if, as is too often the case, sulphuric acid has been used to brighten the colour, or in dyeing, these changes are hastened. This peculiarity, in a greater or less degree, is common to all tans of the "catechol" type; and cassia or quebracho leather is so sensitive to light that it is quite possible to print a photographic negative upon it in a week or two's exposure. The use of such unsuitable leathers is one of the principal causes why leather bindings have fallen into disrepute, but moroccos properly tanned with sumach (or other "pyrogallol" tans) are little sensitive to light, and much less affected by gas, and may be relied on to last for generations if properly cared for[1].

The skins, whether goat or sheep, are usually prepared for tanning by liming (often in "arsenic limes"), puering, and drenching, as described in Chapter X. The tannage is usually in paddle-vats, the ground sumach being added direct to the liquor,

[1] For more detailed information on the preservation of leather see Chapter XXI.

which is freshly made with warm water, as the sumach tannin rapidly ferments and becomes useless, especially in contact with the exhausted leaves. The tannin penetrates rapidly and the tannage lasts at most a few days. "Roans" are sheepskins tanned by a peculiar process, probably of Eastern origin, which only requires about 24 hours. The skins are sewn up, now usually by a sewing machine, into bags much of the form of the animal, turned inside out, and filled with sumach liquor through one of the shanks, which is tied up after introducing a handful of unground leaves (to choke small leaks) and a little air. The skins are allowed to float a few hours in a vat of warm sumach liquor, and then piled on a draining board to press the liquor through them; and again refilled and the process repeated.

After tanning, skins are oiled, dried out and stored, constituting "crust stock," ready to be finished as required; for the colours and finishes demanded are so varied that it is usual to wait for orders.

The first step in finishing is to soak the skins in warm water, and strike them out with a slicker on a sloping table to remove dirt and surplus tan. "Clearing" with a weak bath of sulphuric acid is very common at this stage, and much brightens the colour of the leather, but is apt to lessen durability; and if any free acid is left in, leads to rapid decay.

Formic acid is now often substituted and is much less dangerous, though more costly and not quite so effective. If the dyeing is to be with "basic" coal-tar colours, the washing must be very thorough, as these colours are precipitated by dissolved tannin in the bath and wasted. Instead of complete washing, the tannin is often fixed by a bath of salt of antimony or some other metal, the metallic tannate acting as a mordant for the dye, and so giving a fuller shade. As "basic" colours are very apt to dye unevenly on "weak grain," "acid" dyes (salts of colour-acids) are frequently used. These have less intensity of colour, but dye more evenly and do not precipitate tannin. Acid must be added to the dye bath to set free the colour-acid, and the excessive use of sulphuric acid for this purpose is one of the causes of the decay of many modern leathers. Formic acid may be substituted with good effect, but the weaker organic acids are not efficient. Formerly dye-woods were largely used, but with the exception of logwood for blacks, and a little peachwood and fustic, they have been completely superseded by the synthetic colours.

The old English method of dyeing is to handle or "turn" two or three dozen skins in a bath or "tray" until coloured to shade, the dye being added in two or three successive portions. To economise dye, and keep the flesh-sides white, the skins are generally

9—2

"paired," flesh to flesh, by pressing together with a slicker on the table, and continue to adhere throughout the process. Sometimes, instead of "pairing," they are "pleated," by folding, flesh in, down the back. Where great numbers of skins of one colour are required the paddle-vat is largely used, but in this case the flesh side is also coloured. In the German method, two narrow trays are used, each taking one pair of skins only. The skins are entered in a nearly exhausted liquor, turned a few times, and transferred to the second tray through which only one pair has previously passed. The exhausted tray is now emptied and a new liquor made by adding strong dye solution to hot water, and the skins are turned in this till the required colour is obtained, the whole operation taking only a few minutes. It is thus perhaps as rapid as the English method, since in the latter the skins dye very slowly towards the end of the operation in the nearly exhausted liquor, and 1—1½ hours is often required. In both methods a good deal of dye is wasted, as the bath can only be partially exhausted.

After dyeing the skins are struck out and dried, very frequently nailed on boards to strain and flatten them. The next operation is usually "graining." The skins are slightly damped, laid grain-up on a table, and a portion is folded over, and the crease is drawn across the skin with moderate pressure by a

flat "board" covered with cork, which is held by
passing the hand through a strap across the back.
As the grain is always slacker than the flesh, this
treatment covers it with minute wrinkles.

If the skin is "drawn over" from head to butt, or
vice versa, these wrinkles are nearly parallel, and
produce a "willow grain" (from its resemblance to
willow leaves), which is often seen on bags and
coloured shoes. If the drawing down is repeated
across the skin, a square or "box" grain is formed,
and "morocco grain" is produced by again crossing
both ways from corner to corner. To produce an
even grain on all parts of the skin is highly skilled
work, as the pressure must be exactly adjusted to
the varying thickness and firmness. Even with the
greatest expertness the fineness of the grain is dependent on thickness and texture, and it is impossible
to produce a "bold" or coarse grain on a thin or
hard skin. The "bold-grained seal" so much used
for bags, purses, and pocket books can only be
formed naturally on small, thick and rather soft
skins; and the thinness, if required, is obtained by
splitting the finished skin with the band-knife
machine (p. 124).

CHAPTER XX

OIL LEATHERS

IT has been noted that the use of fats is one of the earliest of primitive methods for the conversion of skin into leather, and suggested that the principal object of the fat was to coat the fibres and prevent their subsequent adhesion. Though this is doubtless true, it does not appear to be the whole truth, for ordinary wash-leather, which is a typical oil leather, can be washed with hot soda solution, which would remove both oils and their oxidation products, and still retains all the qualities of a very perfect leather.

Wash-leather, or as it is often called "chamois" leather, is now always made from the flesh-split, or "lining" of sheepskins, the thin grain of which constitutes the "skiver" mentioned in the previous chapter. The splitting is done on pelts fresh from the limes by a machine with a vibrating knife against which the skin is drawn, and to get sufficient plumpness the liming must be thorough. The skins, either at once, or after a slight drenching, are thrown into a fulling mill or "faller stocks" together with some sawdust, and are kneaded till they reach a semidry and somewhat porous condition, when they are sprinkled with fish oil. The stocking is continued, with occasional pauses to allow the goods to cool, and re-sprinkling with oil at intervals; till the original limey smell disappears and is replaced by a somewhat pungent odour; and the skins are thoroughly saturated with oil. They are then packed in a box and covered up, when they rapidly heat by the oxidation of the oil (and might even take fire, if not taken out in time and hung up or spread on a floor to cool); while at the same time a good deal of very irritating acrolein vapour is evolved. This packing is repeated two or three times till the oxidation is complete, heating no longer takes place, and the skin is yellow throughout and fully leathered. The remaining operations consist in pressing out the surplus oil (dégras or sod oil, which is valuable for currying), washing with warm alkaline solutions to remove the residual oil, and finally drying, and "fluffing" on an emery wheel to produce a smooth and velvety surface.

The French method differs somewhat from the English. Instead of sprinkling with oil in the fulling mill, the skins are shaken out and regularly oiled on a table and folded into bundles before returning to the mill; and between the successive stockings are hung up for 8 or 10 hours in a warm room where considerable oxidation takes place. Mixtures of seal and whale oil are also substituted for pure cod, with the result that the final skins remain of a paler yellow, and the oil is more liquid and less viscid and is probably used in larger quantity. The surplus is recovered after dipping in hot water by wringing, which in former times was done by folding each pair of skins round the fist so that the ends overlapped, and formed a link in a chain of skins of which one end was attached to a fixed hook, and the other to a sort of winch by which it was twisted till the oil was wrung out; hence the name *première torse* for the first and finest *moëllon* or *dégras*. At present, actual wringing is superseded by the hydraulic press. The dégras is used for the finer sorts of currying, but only after mixing with tallow, wool-fat, and other things, which sometimes make it more suitable for its purpose. Genuine dégras usually contains about 25 $^\circ/_\circ$ of water, its ready emulsification with which is one of its most important qualities; and much is now made by the direct oxidation of oils without the intervention of skins. A coarser sort of dégras or sod oil is recovered by acidifying the alkaline washings.

In addition to the use of wash-leather for cleaning and polishing, oil-leather, sometimes actually from deer and antelope skins, is employed for gloves and the like; and for this purpose is often bleached and dyed. The oldest method of bleaching was similar to that employed for textiles, and depended on exposure to air and sunlight in a moist condition, but this is generally superseded or at least supplemented by chemical treatment such as that of soaking in weak solutions of potassium permanganate ("Condy's fluid"), followed by treatment with oxalic acid or bisulphites. Sodium peroxide is also used. When skins which have not been split are oil-dressed, the grain surface is first removed by "frizing" or scraping with a sharp two-handled knife on the beam; and this greatly facilitates the subsequent dress, though not actually essential.

An oil-dressed leather formerly important for military accoutrements is buff-leather, made from ox or cow hides by a process identical with that used for wash-leather, except that the heating must be stopped before the leather becomes too soft and clothy. The hides are subsequently "tucked" or shrunk by dipping in hot soda solutions, and finished by bleaching and fluffing. "Tucking" is also often applied to the lighter glove leathers.

The theory of the oil-process is still somewhat obscure. Knapp's explanation that it consists in coating the fibres with oxidised oil products is negatived because it withstands the action of the alkaline solutions used in its manufacture, which saponify and dissolve oxidised oils. The presence of lime seems essential in ordinary buff-leather, and it is not impossible that the fibres may be coated with lime-soaps, which are not dissolved by alkalies. A tempting explanation founded on the tanning effect of aldehydes was that the effect was due to the acrolein or acryl-aldehyde produced in large quantities from the glycerine of the oils in heating: and though good chamoising can be done by the fatty acids of fish oils from which all glycerine has been removed, it is known that considerable quantities of higher aldehydes are formed from the more oxidisable fatty acids themselves by the breaking of the hydrocarbon chain at the double linkages.

Leathers almost identical with oil-leathers can be made by the direct action of slightly alkaline solutions of formaldehyde on raw hide, as in the Pullman-Payne process, and large quantities of buff-leather have been made in this way for military purposes, and have possibly contributed to its present disuse, since unless extreme care is taken to remove the superfluous formaldehyde, the fibre gradually becomes brittle and tender. Formaldehyde leather is naturally quite white, and such leathers have actually been dyed yellow and then bleached on the surface to imitate the genuine product, which is always yellow internally.

Both aldehyde and bleached chamois leathers can be dyed, though some skill and knowledge are required; but much of the leather which was formerly made for gloves was simply coloured by rubbing in pigments mixed with pipe-clay, and if washed, returned to its original pale yellow.

Beside the regular oil-leathers, there are others mainly used for mechanical purposes, in which fatty matters are the principal tanning agents. Such a leather is Helvetia or Crown leather, made by coating the wet hide with a paste of flour and animal fat, and drumming in a warm drum till the mixture is absorbed. To penetrate a thick hide about three coatings and drummings of eight hours each are required. The various "raw hide" belt and lace-leathers are mostly combinations of this process with varying proportions of vegetable, alum and sometimes sulphur tannage, and, like the original crown leather are distinguished by great toughness and flexibility. A still simpler form of fat-tannage is employed by the Boers in making the Reims or long thongs which they use for many purposes. The unhaired hide is cut in the classical spiral fashion to a long thong which is made into a sort of skein over a cross-bar and a heavy weight is hung on it; grease is repeatedly applied to its moist surface, and is gradually worked in by twisting and untwisting the skein like a roasting-jack, while at intervals its position on the cross-bar is altered so that all parts may be evenly stretched and impregnated.

It need hardly be pointed out that a certain degree of oil or fat tannage is produced in all leathers in the manufacture of which these materials are used.

While even ordinary animal fats will produce a soft and flexible leather, it is only the more oxidisable drying oils which will produce a characteristic oil-leather, and though fish oils are invariably used in practice, Fahrion has shown that a similar effect may be obtained with linseed, and the very tough white Japanese leather used for brace tabs is made by repeated soaking in water, oiling with rape oil and exposure to the sun. The Nappa leather of the United States is a sort of oil-leather made by fulling with soft-soap, but the German nappa is a quite different material.

CHAPTER XXI

THE USE AND CARE OF LEATHER

It seems fitting to conclude this little book with a few practical hints on the treatment of a material so commonly, and often so unwisely, used as leather. Leather is not only important for boots and shoes, book-binding and upholstery, but in engineering it answers a great variety of purposes, and its applications in harness, saddles and coach-work must not be overlooked.

The leather in boots may be divided into two very distinct classes—the heavy and solid leather of the sole; and the soft and flexible material of the upper parts of the boot. The sole has to stand the severest wear, exposed as it is to constant friction both in wet and dry condition, and on its quality both the health and comfort of the wearer largely depend. It is, however, impossible for the public to judge of quality once the leather is made up into a boot, and the only useful advice which can be given is to go to a reliable dealer and to pay a fair price. Manufacturers do not work for philanthropy, and a cheap boot, however smart in finish, is not worth more than its price. In any case labour and capital charges form so large a part of the cost that the cheapening is mainly effected by the use of inferior material. It is also very poor economy to wear a single pair of boots continuously, and especially so in wet weather. Either the soles are never properly dried, and so injurious to health, and from their soft condition subject to rapid mechanical wear, or they are dried much too rapidly, and the leather is injured by heat, and curled up by uneven drying, so that one layer of the sole parts from another and water gets in, not only through the leather but at its edges, and the boot is rapidly ruined. The tanner takes at least a week to dry his single thickness of leather, and it is not to be expected that the domestic servant can do it

satisfactorily in twelve hours with much more primitive
appliances. Few people realise the sensitiveness of
wet leather to heat; a pair of wet soles rested on a
railway hot water tin for half an hour will be absolutely burnt, so that when dry they break like glass,
and the wearers usually make claims on the shoe-
maker! The utmost temperature wet leather will
stand is about 120° Fah., or as hot as the hand will
bear, and the worse the leather the more sensitive
it is. Dry leather will stand much higher temperatures; and before waterproofing leather in baths of
melted waxes the manufacturer dries it with extreme
thoroughness. No waterproofing method can be very
satisfactorily applied by the amateur to the finished
boot. Probably painting the sole when dry with
boiled linseed oil, or a mixture of this and oak
varnish, is one of the best. Some of the celluloid
(nitro-cellulose) varnishes may also be suitable; and
melted paraffin wax may be applied with a brush to
the thoroughly dry and warm boot without much
danger, and melted in by cautious holding to the
fire.

The upper-leathers of boots also demand proper
care, but in these days the variety of tannage is so
great that it is difficult to give general directions.
Boots which have been soaked should be dried slowly
and evenly, and preferably upon "trees," so as to
retain their proper shape; and few sorts of leather

would not be better under these circumstances for
a light oiling, which should be applied while the
leather is wet. Castor-oil of the common sort sold
for lubrication is one of the best, and will not prevent polishing; but almost any non-drying oil will
do good to the leather, and even raw linseed oil may
be moderately used. "Dubbing," properly a mixture
of fish oil and tallow, is not specially suitable for
domestic application to boots, except perhaps for
heavy shooting boots, for which any soft grease will
do. Box-calf and other chrome leathers will stand a
good deal of heat, and dry soft. They are not injured
by oils, though if too freely oiled they lose their
gloss; and even a little oil will darken and spoil the
appearance of coloured boots; but all sorts of chrome
leathers are very sensitive to the old-fashioned
blackings, which cause them to crack on the surface.
Most modern shoe-pastes are mixtures of waxes and
turpentine, or of wax soaps, and, moderately applied,
are not harmful. No leathers can be exposed to the
fumes of burnt gas without injury, and the writer has
seen a pair of heavy shooting boots completely ruined
by being left on an upper shelf in a kitchen where
much gas was burnt.

Gas fumes are also the cause of great injury to
book-bindings and leather-covered furniture, owing
to the sulphuric acid produced from the combustion
of the small quantities of sulphur always present;

and leathers tanned with catechol tannins, such as the so-called "Persian" goat and sheep (tanned in India with cassia bark), are particularly sensitive, while genuine sumach moroccos are much less so. Even when not exposed to gas fumes, "Persians" are comparatively short lived, and though quite good for slippers and other things which will be worn out in a year or two, should never be used for bindings or furniture. Many leathers are ultimately ruined by the careless use of sulphuric acid in manufacture, but since the publication of the report of the Society of Arts Commonwealth on the subject in 1905 this has received a good deal of attention, and it is now easy to obtain leathers guaranteed by the manufacturers to be quite free from mineral acids. The form of damage most prevalent in the many libraries visited by the Committee was that of "red decay," in which the surface crumbled, exposing the reddened leather below, which in bad cases was completely disintegrated and fell to powder. This defect is largely due to the peculiar character of the catechol tannages (see p. 129), but is much aggravated by acidity. Sunlight also has a most injurious effect of the same kind; and the windows of libraries where valuable books are kept should be well protected with blinds, or glazed with pale yellow or greenish glass; and leather-covered furniture should also be kept as much as possible from direct sunshine. Even under favourable circumstances "Persians" cannot be expected to last more than 15 or 20 years, while sumach and oak-bark tanned leathers are almost immune from decay. It was specially noticed in the public libraries visited that the books most handled were often the best preserved, from the slight traces of grease they received, and there is no doubt that an occasional rub with a cloth very slightly oily with lanoline, vaseline, or oil would tend to the preservation of the leather. Colourless shoe-pastes may also be used on bindings with good effect. For further information on these points the reader is referred to the full Report of the Society of Arts[1], and to *Leather for Libraries*[2], by Dr J. G. Parker and others.

The choice of leathers for bindings is largely a matter of taste, but modern sumach morocco seems more reliable than calf, though many old calf bindings have lasted for centuries. Pigskin and seal morocco are also good, and for cheaper work suitably tanned sheepskin, especially that similar to the "roller leather" of the cotton spinners, is quite reliable. The thin "skivers" or sheepskin grains may be quite permanent as regards decay, but naturally will not resist much mechanical wear. Vellum or parchment

[1] Bell and Sons, London, 1905, for Society of Arts.
[2] Library Supply Co., Queen Victoria St., E.C., 1905, for Library Assoc.

is very lasting, but apt to crack at the joints. For upholstery, goat, seal, and pigskin are all good, and some
very beautiful leathers, hardly to be distinguished from
morocco, are now made of split hide, which has the
advantage of large surface without joints. Similar
leather is also generally used for motor body-work,
and for this purpose is waterproofed with a thin
celluloid or nitro-cellulose coating, such as is used
for leather cloths.

Bag and portmanteau leathers are usually split hide,
and for so-called "solid" leather articles, are stiffened
with millboard or strawboard; but genuine solid
leather sufficiently thick to be stiff enough without
support, though more costly, is much more durable.
The cheaper suit cases and portmanteaus are often
only covered with tanned sheepskin (bazils), or at
best with the inferior parts of split hide. If soap
is used for cleaning such articles it should be as
neutral (free from alkali) as possible, and after its
use the leather, while still damp, should be rubbed
with a rag slightly oiled with linseed or some
other oil.

The mechanical uses of leather are too numerous
to mention in detail. In the house, most screw
water-cocks have leather washers, and for this purpose chrome-tanned is preferable to vegetable-tanned
leather, and may even be used for hot water in place
of rubber or "vulcanised fibre."

For durability in machine-belting no material has been found superior to leather, though for some purposes woven fabrics give excellent service. Where great flexibility is important, as in driving high speed machinery with small pulleys, chrome leather, "raw hide," and some of the combination tannages are superior to vegetable-tanned, but are very liable to damage at the edges if run through forks, and sometimes troublesome in stretching and getting out of shape. All belts are better for occasional washing with a suitable soap and warm water and oiling while damp with a moderate quantity of castor, olive or tallow oil. Mineral oil, if free from acid, is not injurious to leather, but has not the same permanent softening effect as the oils mentioned. Rosin is injurious, and tends to make leather crack. For heavy drives, breadth is to be preferred, as far as possible, to thickness.

An important mechanical use of leather is for cup and ring leathers for pumps and hydraulic rams, which are pressed into shape in dies in a moist state, allowed to dry, often trimmed on a lathe, and afterwards soaked in hot oil.

Thin leathers are used to cover the drawing rollers in spinning machinery, sheep for cotton and calf for heavier threads, and a speciality in textile industries is the leather for carding machines, covered with a sort of brush of fine steel wire. "Pickers" are blocks of compressed raw hide used to stop and throw the shuttle in power-looms, and the "picker band" is a very flexible and tough thong by which the picker is actuated.

BIBLIOGRAPHY

Leather Industries Laboratory Book. H. R. Procter. E. & F. N. Spon, Ltd., London.
Principles of Leather Manufacture. H. R. Procter. E. & F. N. Spon, Ltd., London.
Leather Chemists' Pocket Book. H. R. Procter, E. Stiasny and H. Brumwell. E. & F. N. Spon, Ltd., London.
The Manufacture of Leather. H. G. Bennett. Constable & Co., London.
Puering, Bating and Drenching. J. T. Wood. E. & F. N. Spon, Ltd., London.
The Sheep and its Skin. A. Seymour-Jones. Leather Trades Review, London.
Leather Trades Chemistry. S. R. Trotman. C. Griffin & Co., Ltd., Strand, London.
The Manufacture of Leather. C. T. Davies. Sampson Low, Marston & Co., London.
Modern American Tanning. Fleming. Jacobsen Publishing Co., Chicago.
Handbuch der Chromgerbung. J. Jettmar. Schulze & Co., Leipzig.
La Tannerie. L. Meunier & C. Vaney. Gauthier-Villars, Paris.
Das Färben des lohgaren Leders. J. Jettmar. Verlag von Bernh. Fredr. Doigt, Leipzig.

BIBLIOGRAPHY 149

Praxis und Theorie der Leder-erzeugung. J. Jettmar. Julius Springer, Berlin.

Hides and Skins. Shoe & Leather Weekly, Chicago, U.S.A.

Die Chromgerbung. J. Borgman. M. Krayn, Berlin, W.

Die Feinlederfabrikation. J. Borgman. M. Krayn, Berlin, W.

Die Rotlederfabrikation. J. Borgman. M. Krayn, Berlin, W.

Leather Manufacture. Watt. Crosby Lockwood & Son, London.

The Leather Worker's Manual. Standage. Scott, Greenwood & Co., London.

Traité Pratique de la Fabrication des Cuirs et du Travail des Peaux. A. M. Villon, U. J. Thuau. Ch. Beranger, Paris.

Leather Dressing, Dyeing, Staining and Finishing. M. C. Lamb. Leather Trades Publishing Co., London, S.E.

Leder Industrie. W. Eitner. Faesy & Frick, Wien.

Leather for Bookbinding. Society of Arts. G. Bell & Sons, London.

INDEX

Abies canadensis, 122
Acid, amino-acetic, 20
  boracic, 45
  carbolic, 100
  dyes, 97
  formic, 24, 74
  hydrochloric, 74
  lactic, 74
  soaking, 23
  stearic, 69
  sulphuric, 48
  sulphurous, 24
  swelling, 42
Acids, "strong," 42
  table of "strength" of, 47
  "weak," 43
Adsorption, physical, 76
Aescherung, 30
Ageing, 85
Alkaline soaking, 23
  swelling, 28
Alum, 80
Aluminium sulphate, 81
Amine, 20
Amino-acetic acid, 20
  groups, 82
Ammonium chloride, 50, 66
Amœba, 9
Amphoteric, 21
Arsenic sulphide, 32
Atoms, 17

Bacillus coli, 63
  erodiens, 63, 66
Bacteria, 52
Bacterium furfuris, 59
Band knife machine, 124
Basic chrome process, 88
  dyes, 97
  salts, 81
  salts, formation of, 82
Bating, 56
Becker, 63, 66
Belt laces, 84
Bends, 37
Blacksmith's aprons, 84
Bleaching extracts, 113
Bloom, 102
Boracic acid, 45
Boarding, 127
Böttger, 32
Buffalo method of unhairing, 31
Buffing, 127
Butt, 37, 105
  rolling machine, 116

Cæsalpinia coriaria, 102
Calf-kid, 86
Carbolic acid, 100
Cassia auriculata, 129
Castanea vesca, 102
Catechol, 100
  tannins, 102

Caustic soda, 24
Celluloid varnish, 128
Cementing substance, 16
Chamois leather, 134
Chestnut, 102
Clark's water softening process, 41
Collagen, 21
Colloid jelly, 21
Colloidal solutions, 71, 103
Condy's fluid, 137
Connective tissue, 16
Corium, 12, 67
Crown leather, 87, 139
Crust stock, 130
Curriers' shaving knife, 123
  steels, 123
Currying, 121

Davy, Sir Humphrey, 70
Dégras, 136
Deliming, 37
Distilled stearine, 126
Divi-divi, 102
Drawn grain, 109
Drenching, 56
Dried hides, 7
Drum-stuffing, 126
Drying of leather, 114
Drying oils, 125
Drysalted hides, 8
Dubbing, 125, 143
Dyes : coal-tar, 97
Dye woods, 131

Egg-yolks, 85, 87
Ellagic acid, 101
Enzymes, 29
Epidermis, 12
Erectores pili, 15
Erodin, 66
Equilibria, 70
Equilibrium, quadruple, 78

Fahrion, 140
Fat-liquoring, 91, 93
Fermentation, 52
Flagella, 53
Fleshing, 35
  knife, 34
  machine, 38, 39
Flour, 85
Formaldehyde leather, 138
Formic acid, 24, 74
Frizing, 137

Gambier, 103
Gelatine, 22
  swelling of by acids, 74
  tannate, 70
Glacé kid, 86
Glassing, 127
Glassy layer, 15
Glazed kid, 99
Glazing machine, 98
Gloves, kid, 85
Glucose, 90
Graining, 132
Guncotton, 70

Hair-papilla, 13
  sheath, 13
Handlers, 109
Hauling, 27
Head-knife, 122
Heinzerling, 93
Helvetia leather, 87, 139
Hemlock, 122
Hide-mill, 23
  salting, 6
Hurdle, 85
Hyaline layer, 15
Hydraulic leathers, 147
Hydrochloric acid, 74
Hydrolysis, 40
  equilibrium, 82

"Ions," 44

Japanese leather, 140

Keratin, 12, 21
Kip, 5, 118
Knapp, 68, 69, 88, 137

Lactic acid, 74
Layers, 109
Liming, 26
Linder and Picton, 71
Logwood, 102

Martin Dennis, 88
Mellow liquors, 106
Meunier, 68
Mimosa, 103
Moellon, 136
Molecules, 17
Moon-knife, 86
Moors, 80
Moroccos, 97
Mucous layer, 13
Myrobalans, 102

Nappa leather, 140
Neutralisation of Chrome leather,
    91
Non-tannins, 106
Nucleus, 9

Oak-bark, 101
    galls, 102
Oakwood, 101
Oleostearin, 126
Ostwald's hydrolysis formula, 76

Paddle-vat, 57
Pancreatin, 65
Pancreas extract, 66
Paraffin wax, 126
Parchment, 68

Parker, Dr J. G., 145
Pathological tannins, 102
Payne and Pullman, 30
Pebbled grain, 127
Pelt, 68
Pepsin, 65
Perching, 86
Persians, 128
Phlobaphenes, 101
Phloroglucol, 100
Picker bands, 148
Pickling, 72
Picton, 71
Pinholes, 58
Pistacio lentiscus, 102
Polyp, 11
Potassium carbonate, 68
Première torse, 136
Printing, 127
Proteids, 20
Protoplasm, 9
Pseudopodia, 9
Puering, 56
Pullman, 30
Pullman-Payne process, 138
Putrefaction-bacteria, 23
Putting out, 93
Pyrogallol, 100
    tannins, 102

Quadruple equilibrium, 78
Quebracho, 103
Quercus infectoria, 102
    pedunculata, 102
    robur, 102
    sessiliflora, 102

Raw hide, 139
Realgar, 85
Red decay, 144
"Reds," 101
Reims, 139
Rhus coriaria, 102

Roans, 130
Röhm, Dr, 66
Rounding, 37, 105

Salomon, W. J., 65
Salt, common, 72
Salt-stains, 7
Scudding, 35
Schultz, 88, 93, 94, 95
Seasoning, 98
Sebaceous glands, 14
Setting, 27
Sharpening limes, 30, 105
Sizing, 127
Skin-chloride, 79
Skivers, 145
Slicker, 122
Soaking, 22
Sodium acetate, 49
  formate, 78
  sulphide, 24
  thiosulphate, 94
Sod oil, 126
Solutions, colloidal, 71
Splitting machine, 124
Spueing, 125
Staking, 73, 84
Staling, 26
Stearic acid, 69
Stiasny, 50, 92, 104
Striking, 115
  pin, 114
"Strong" acids, 42
Structural formula, 18
Stocks, 23
Stuffing, 125
Sudoriferous glands, 15
Suède, 86
Sulphur-tannage, 95

Sulphurous acid, 24
Sumach, 102
Suspenders, 105
Sweat glands, 15
Sweating, 26, 108
Swelling alkaline, 28
Synthetic tannins, 104

Table grease, 125
Table of "strength" of acids, 47
Tannate of gelatine, 70
Tanners' beam, 34
Tannic acids, 100
Tawing, 81
Temporary hardness of water, 41
Tray dyeing, 131
Tucking, 137

Uncaria gambia, 103
Unhairing, 33
  knife, 34
  machine, 36
Unorganised ferments, 53

Valencies, 18
Valonia, 102
Vatting, 113
Vellum, 68

Wash-wheel, 22
Wattle, 103
Water softening, Clark's process
  of, 41
"Weak" acids, 43
Weak-grain, 25
Whip lashes, 84
Whitening, 126
Wilson's striking machine, 116
Wood, J. T., 47, 58, 65

For EU product safety concerns, contact us at Calle de José Abascal, 56–1°, 28003 Madrid, Spain or eugpsr@cambridge.org.